Environmental Impacts of Artificial Ice Nucleating Agents

Environmental Impacts of Artificial Ice Nucleating Agents

Edited by

Donald A. Klein
Colorado State University

Dowden, Hutchinson & Ross, Inc.
STROUDSBURG, PENNSYLVANIA

This book was prepared with the support of NSF Grant ENV73-07821. However, any opinions, findings, conclusions and/or recommendations herein are those of the authors and do not necessarily reflect the views of NSF.

Library of Congress Catalog Card Number: 78–7985
ISBN: 0–87933–334–0

80 79 78 1 2 3 4 5
Manufactured in the United States of America

LIBRARY OF CONGRESS CATALOGING IN PUBLICATION DATA

Main entry under title:
Environmental impacts of artificial ice nucleating agents.
Bibliography: p.
Includes index.
1. Silver—Environmental aspects—Congresses.
2. Silver compounds—Environmental-aspects—Congresses.
3. Rain-making—Environmental aspects—Congresses.
I. Klein, Donald A., 1935-
QH545.S55E57 574.5′222 78–7985
ISBN 0–87933–334–0

Distributed world wide by Academic Press,
a subsidiary of Harcourt Brace Jovanovitch,
Publishers.

Contents

Environmental Impacts of Artificial Ice Nucleating Agents

INTRODUCTION

This volume provides an analysis of available information and the scientific conclusions from a National Science Foundation-sponsored workshop on the possible environmental impacts of nucleating agents used in weather modification programs. The workshop was held November 10-11, 1976 at Vail, Colorado, in conjunction with a U.S. Bureau of Reclamation Skywater meeting, which allowed workshop members to interact with individuals concerned with more general environmental aspects of weather modification programs.

This is not a verbatim transcript of the meeting discussions; the meeting was used to assure that the general conclusions given here reflect the best possible assessment of current knowledge.

The field of weather modification has continued to develop since the 1946 to 1950 period, and this technology has been used for hail suppression, winter and summer precipitation augmentation, fog suppression, tornado and hurricane modification, and lightning abatement, as specific examples. Weather modification, both intentional and inadvertant, has also led to legal, economic, sociological, and environmental questions.

One question about weather modification technology concerns possible environmental impacts of nucleating agents. Since the beginning of the modern era of intentional weather modification in 1946, a wide range of nucleating agents have been used, however, the major nucleating agent has been silver iodide. There have been several recent assessments of weather modification environmental impacts. Cooper and Jolly (1969, 1970) provided an analysis of potential environmental impacts of weather modification programs, including nucleating agents, based on available literature drawn primarily from general medical and environmental research areas.

Waggoner (1966) and Whittaker (1967) assessed general climatological aspects of weather modification, however, nucleating agent impacts were not considered. In a general review of silver environmental impacts, Smith and Carson (1977) considered possible silver contributions from weather modification to the total silver added to the environment.

This symposium and summary volume were designed to meet the following objectives: (1) to assess the scientific information presently available, (2) to summarize our current knowledge in each area, which could be useful to program managers and to environmental and legal agencies, (3) to recommend areas requiring additional work, and perhaps more important, to note areas where sufficient information is already available, and (4) to provide recommendations on operational aspects of environmental monitoring of nucleating agents, which might apply to the management of future weather modification programs. With the primary use of silver compounds as nucleating agents, this will be the major thrust of the book, although in a later chapter other agents, including organics and inorganics, will be discussed.

The following individuals were members of the workshop, and those contributing chapters to the summary volume are noted with an asterisk:

Duane Anderson
College of Pharmacy
University of Wyoming
Laramie

James A. Bailey
Fishery and Wildlife Biology
Colorado State University
Fort Collins

Richard Barchett*
Department of Meteorology
University of Wisconsin
Madison

Keith Brown
North American Weather Consultants
Goleta, California

Bonnie Carson
Midwest Research Institute
425 Volker Boulevard
Kansas City, Missouri

Patrick H. Davies*
Colorado Division of Wildlife
Fort Collins

William Finnegan
Naval Weapons Center
China Lake, California

Joel Fletcher
Utah Water Research Laboratory
Utah State University
Logan

Paul M. Giordano
National Fertilizer Development Center
Tennessee Valley Authority
Muscle Shoals, Alabama

Donald G. Girvin*
Lawrence Berkeley Laboratory
Berkeley, California

John Goettl*
Colorado Division of Wildlife
Fort Collins

Donald A. Klein*
Department of Microbiology
Colorado State University
Fort Collins

Harold Koontz
Biological Sciences
University of Connecticut
Storrs

Willard Lindsay*
Department of Agronomy
Colorado State University
Fort Collins

Gerald Mulvey*
Department of Atmospheric Sciences
Colorado State University
Fort Collins

Bernard Vonnegut
Department of Atmospheric Sciences
State University of New York
Albany

Tad Weaver
Department of Botany
Montana State University
Bozeman

In addition, the following individuals contributed to the discussions:

Don Brosky
Kodak
Windsor, Colorado

Richard Dirks
National Science Foundation
Washington, D.C.

Richard Eddy
U.S. Bureau of Reclamation
Denver, Colorado

Wallace Howell
U.S. Bureau of Reclamation
Denver, Colorado

R. F. Reinking
NOAA
Boulder, Colorado

The following individuals prepared chapters for inclusion in the summary volume:

James W. Ball
U.S. Geological Survey
Menlo Park, California

Jo Matter Burchard
U.S. Geological Survey
Menlo Park, California

Everett A. Jenne
U.S. Geological Survey
Menlo Park, California

M. Sadiq
Department of Agronomy
Colorado State University
Fort Collins

During final preparation of the manuscript, Gayle Robison provided valuable assistance in drafting chapters 8, 9, and 12. Gail Ellis, Susan Martino, and Susan Taylor prepared the bibliographic materials, index and layout of the book. Leila Menges and Diane Fox of the Natural Resource Ecology Laboratory, Colorado State University, supervised the final typing and phototypesetting, which was completed by B. Vader Phototypesetting, Fort Collins, Colorado.

The following individuals and companies have given permission to reproduce copyrighted materials:

Academic Press, Inc. (Figs. 2-1, 9-7, 13-2); American Association for Cancer Research (Table 13-3); American Chemical Society (Figs. 5-3, 5-4); American Meteorological Society (Table 7-3; Figs. 7-4, 7-5, 7-6); Cambridge University Press (Figs. 9-4, 9-5, 13-3); D. Reidel Publishing Company (Fig. 9-3); Elsevier/North-Holland Biomedical Press (Tables 9-2, 12-1, 12-2); Geological Survey of Canada (Table 7-1); International Atomic Energy Agency (Figs. 9-1, 9-2); John Wiley & Sons, Inc. (Fig. 5-1); Journal of Pharmacy and Pharmacology (Fig. 6-3); N. L. Kent (Table 8-9); Macmillan Journals Limited (Fig. 6-4); Midwest Research Institute (Table 1-1); New York Academy of Sciences (Table 13-1); Pergamon Press, Inc. (Fig. 6-2; Table 9-1); Plenum Publishing Corporation (Fig. 5-2); The Regents of the University of California (Fig. 13-1); The Royal Netherlands Society of Agricultural Sciences (Fig. 8-1); The Royal Society of Britain (Fig. 8-4); Springer-Verlag, Inc. (Table 6-1); University of Missouri (Table 9-3); Arthur Wallace (Table 8-1); T. Weaver (Table 8-2); The Wistar Press (Tables 13-2, 13-4); World Meteorological Organization (Table 11-1).

This volume was prepared with support from the National Science Foundation under Grant ENV 73-07821-A04. The opinions, findings, and conclusions or recommendations are those of the workshop members and chapter contributors, and do not reflect the views of the supporting organization. The continuing support of R. A. Dirks and C. S. Downie of the National Science Foundation, and of A. M. Kahan and W. E. Howell of the U.S. Bureau of Reclamation is gratefully acknowledged.

Donald A. Klein
Department of Microbiology
Colorado State University
Fort Collins, Colorado

1

Amounts of Silver-Derived Agents Used in Weather Modification in Relation to Contributions from Other Sources

D. A. Klein and G. Mulvey

Mankind has had a long and continuous exposure to silver and its various compounds because of its use for decoration, utensils, and coinage, as well as its unrecognized use as an inhibitor of water-borne pathogenic microorganisms, as summarized by Just and Szniolis (1936) and Hosenfeld (1938). Information on silver release to the United States environment has been summarized by Smith and Carson (1977), and data on silver levels in lake and ocean bottom sediments has been collected, as by Bruland et al. (1974) for the southern California area.

Compared to silver releases from mining, industrial, and photo processing activities, silver used as a nucleating agent in weather modification programs contributes very little to total environmental silver levels. According to Smith and Carson's summary of silver in the environment, approximately 100,000 troy ounces or 3.1×10^6 grams of silver is used in weather modification programs per year (Table 1-1), which is 0.15% of the total silver estimated to be released. Of interest was the estimate of photographic industry losses to air (2×10^6 g), water (1.1×10^8 g), and land (3.3×10^8 g); the latter two rates being markedly higher than those of weather modification programs. The release of silver not associated with weather modification contributes to background aerosol concentrations of approximately 0.01 ng/m^3, and up to 7.0 ng/m^3 over source regions such as urban centers (Mulvey, 1977).

After the basic technology for these programs was developed in the late 1940's, an estimated 10% of the U.S. land area was seeded during the early 1950's (Fleagle et al., 1974). Following this initial period of intensive

TABLE 1-1. *Release of silver to the environment from various sources in millions of troy ounces (Smith and Carson, 1977).*[a]

Loss Category	Total Loss
Mining and Milling	3.1
Primary Smelting and Refining[b]	5.5
Fabrication, Total	0.15-0.26
Use and Disposal, Total	39.1
Cloud Seeding	0.1
Inadvertant Sources	26.8-28.2
Total Release	78.7-84.2

[a]One troy ounce = 31.19 g.

[b]No estimate was given for the amounts released by secondary smelting and refining processes.

activity, weather modification was conducted on a more selective basis, resulting in less material being released to the environment. But the current agent use rate in the U.S. is increasing and even approaching the 1950 maximum (L. O. Grant, personal communication).

The amounts of materials used in recent years in specific programs are available. During 1965 to 1970, in southern France, silver iodide was used at a rate of 4 tons per year (3.6×10^6 g) (Dessens and Lacaux, 1972). This hail suppression project covered an area of 7,000,000 hectares, using 350 ground-based generators during 1970.

A total of 320 kilograms of silver iodide (containing 45.95% by weight silver) were released from the ground during 1970 to 1975 in the Colorado River Basin Pilot Project (Elliot et al., 1976) over a target area of 3,400 km^2, of which 2,657 km^2 was above 2,000 M elevation. The yearly release rate varied from a low of 31 kg to a high of 95 kg per season.

Amounts of seeding agents used in weather modification activities have been reported and monitored by the National Science Foundation and the National Oceanic and Atmospheric Administration (NOAA). Other summaries of weather modification activities have been provided by Charak and DiGiulian (1974) and Charak (1975). These include information on weather modification activities, operators, total target areas, agents and dispensing rates. For the period November 1, 1972 to December 31, 1973, for example, 67 projects were reported, 65 of which were nonfederally sponsored. The programs were operated in 19 states, involving less than 4% of the total land area for the designated target areas (134,000 sq. miles). Silver iodide, dry ice, water droplets, polyelectrolytes (polymeric materials having an electrical charge), propane, and combinations of silver iodide with other ingredients were used. Silver iodide was used in 50 projects, dry ice in 14, and polyelectrolytes were used in 5 projects. Dispensing rates for dry ice and polyelectrolytes ranged from 11 to 1,300 kilograms per hour.

Other seeding agents have also been developed, and some are being used on an operational basis. The use of silver iodide- silver bromide mixtures has been suggested (Vonnegut and Chessin, 1971; Chessin and Vonnegut, 1971). In addition, Passerelli et al. (1973) found that the use of 30% CuI in a CuI-AgI mixture allowed the formation of a nucleant more closely approaching the bond distance spacing of water. This is of interest in relation to the unique set of chemical (and biological) similarities between these two metals, which will be discussed in subsequent chapters. Corrin and Barnes (1976) suggested the use of silver iodide-coated titanium dioxide particles, which could result in silver iodide used at approximately one-tenth the normal rates. Lead iodide is also used extensively as a seeding material in the USSR and some South American countries.

The use of dry ice as a nucleating agent is currently receiving additional emphasis. Schaefer (1976) has noted that dry ice offers economy and superior nucleating effectiveness at all temperatures below 0°C, in comparison with silver iodide. Super et al. (1977) discussed the practicability of seeding convective clouds with dry ice. Dry ice produced at least 2 to 5 $\times 10^{11}$ crystals per gram in super-cooled clouds, and was considered "a practicable and attractive agent for on-top seeding of cumulus clouds." Additional information about nucleating agents other than silver iodide appears in Chapter 14.

Several economic factors might influence the continued use of silver iodide as a nucleating agent. A dramatic rise in the cost of silver consumed in cloud seeding would provide an incentive for reducing this consumption, either by using a coated substrate as suggested by Corrin and Barnes (1976) or by substituting other agents for silver iodide. These considerations will limit the rate of silver dispersion for cloud seeding to perhaps ten times the present rate.

CONCLUSIONS

Since the weather modification profession has confidence in the use of silver iodide as a nucleating agent, it will probably continue to be used in such programs, possibly supplemented by use of agents such as dry ice, which is increasingly being considered. With changes in cost-benefit relationships, other nucleating agents can be expected to supplement silver iodide, depending on their effectiveness, ease of use, and comparative overall delivery costs.

The relative contribution of silver from nucleating agents to environmental silver releases is approximately 0.15%, according to available estimates. In spite of the minor amount released by weather modification programs, this activity's visibility and the unique relationship between individuals and atmospheric processes may lead to greater concern about releases from this source.

2

Atmospheric Transport Processes

G. Mulvey

2.1 INTRODUCTION

The first of many processes controlling silver iodide (cloud seeding material) movement in the biosphere is atmospheric transport. This process, which involves dispersion, transport, and deposition, is of key importance in assessing the implications of the addition of silver iodide to the environment. The location and aerial extent, as well as the timing and amount of silver iodide entering the snow pack, soil, or aquatic environment are controlled by transport mechanisms.

Measurements of silver iodide transport, diffusion, and deposition are rarely made, but measurements of trace silver are more frequently available. The discussion in this chapter, therefore, will be confined to the atmospheric movement of silver from seeding generator operation.

In discussing the transport and deposition of silver particles in the biosphere, it is helpful to first consider silver aerosol sources other than weather modification, which will allow the problem to be viewed in perspective.

2.2 BACKGROUND AEROSOLS

Aerosols are constantly being introduced and removed from the atmosphere by a variety of mechanisms, including sea surface to air transport, gas to particle conversion, wind erosion, anthropogenic activities, volcanoes, forest fires, and the descent of meteoric debris. A relative production rate ranking of these sources is given in Table 2-1.

Sea to air transport, operative over 70% of the earth's surface, is the most important aerosol production mechanism. Primarily produced by the agitation of the ocean's surface, these aerosols are distributed in a way which may be characterized by the equation:

TABLE 2-1. *Relative importance of aerosol production mechanisms* in terms of mass produced/hour.*

Sea to Air Transport of Salts	100
Natural Gas to Particle Conversion	74
Wind Blown Dust	50
Man-Made Gas to Particle Conversion	38
Man-Made Particles	10
Volcanoes	4
Forest Fires	3
Meteoric Debris	1

*Derived from Peterson and Junge, 1971.

$$\frac{dN}{d(\log r)} = C_{M} r^{-\beta_M} \quad \text{for } r > 0.1\ \mu m$$

The importance of this mechanism diminishes as one proceeds inland. The distribution gradually transforms to a continental aerosol which can also be characterized as a power distribution of the form:

$$\frac{dN}{d(\log r)} = C_c r^{-\beta_c} \quad \text{for } r > 0.1\ \mu m$$

where $C_C \geq C_M$
$\beta_C \geq \beta_M$
$\beta_C \cong 3$

For both maritime and continental aerosol distributions, the number concentration usually decreases for particles with diameters less than 0.1 μm. Figure 2.1 depicts a typical aerosol distribution.

One of the major natural sources of aerosols over the continents is the mechanical grinding and dispersion of exposed soils and rocks, some of which contain silver. The aerosol generation from surface material starts as a saltation movement process, generating smaller particles which can then become airborne. One can estimate, from the assumed aerosol size distribution, that the majority of the mass of suspended aerosols are particles between 0.2 and 20 μm in radius. This is supported by the observations of Junge (1963) as shown in Figure 2.1.

2.3. BACKGROUND SILVER

Short- and long-term measurements of silver aerosol concentrations are limited. One long-term measurement series was conducted in

Cleveland, Ohio, at 16 sampling sites over a one-year period during 1971-1972 (King et al., 1976). The average silver concentration was reported to be 1.1 ($\pm$ 0.7) ng m^{-3}. Short-term aerosol studies in urban environments (Brier and Grant, 1973; Harrison et al., 1971; John et al., 1973; Demuynck et al., 1976; Grosch and Georgii, 1976) have yielded estimates of urban silver concentrations in the range from <0.018 to 7.0 ng m^{-3}. Limited sampling of upper level air between 3,800 and 4,300 m MSL was conducted during January 1976, south of Denver, Colorado (Mulvey, 1977), which indicated silver concentrations generally below 0.1 ng m^{-3}, with some values as high as 0.8 ng m^{-3}. The lower range of silver concentrations (<0.1 ng m^{-3}) probably represents the value of background concentrations while the upper range represents concentrations in highly polluted urban environments.

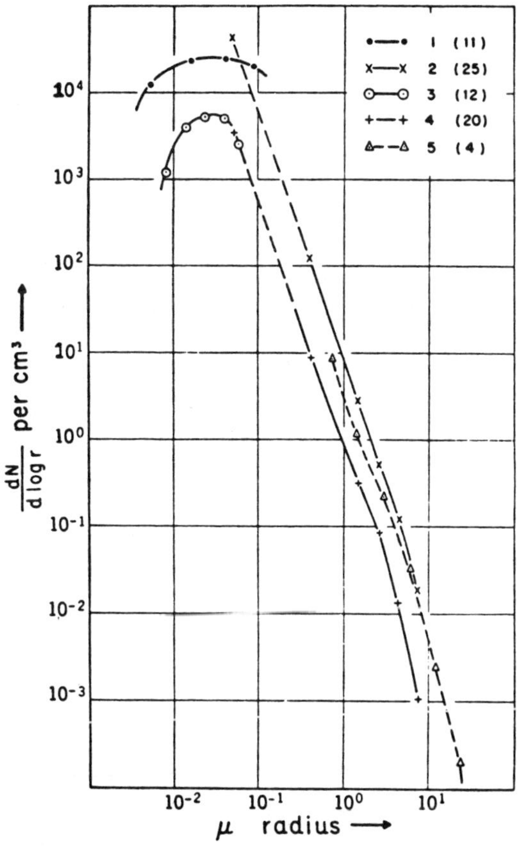

FIGURE 2-1. *Complete size distributions of natural aerosols, average data (Junge, 1963).*

Background measurements of aerosol silver concentrations were also made in relatively clean environments on a short term basis by Mulvey (1977), John et al. (1973), and Brier and Grant (1973). These measurements showed a range of values similar to that of urban environments (0.011 to 7.2 ng m^{-3}).

Estimates of natural aerosol silver can easily be made for the largest continental aerosol source, the saltation process. Since there is no reason to suspect that trace silver particles will preferentially be distributed into a certain size range, we may assume that they are uniformly distributed throughout the various particle sizes. Long-term trends in suspended particulate mass from 20 non-urban sites in the continental United States have shown that an annual average of 28.5 μg m^{-3} are suspended. Estimates of the fraction of silver in the earth's crust and the measured background aerosol mass concentration can be used to calculate the natural background silver concentration in air. The fraction of silver in the earth's crust was estimated to be 0.07 $\mu g/g$ (Mason, 1966). The background silver mass concentration can thus be estimated at 0.002 ng m^{-3}.

Other natural aerosol sources include gas-to-particle conversion, volcanic activity, and sea surface transport. Aerosols formed by gas-to-particle conversion are not likely to contain silver unless it is obtained via particle aggregation. The contribution to the silver budget or cycle of the atmosphere from volcanic activity and meteoritic debris fallout has not been estimated because of lack of silver measurements for these aerosols.

Based on the work of Warburton and Young (1972), particles formed via sea surface to air transport do not seem to contribute a large quantity of silver to the atmospheric aerosol composition. Observation of the elemental aerosol composition for particles (Allee et al., 1976) were used to estimate that about 45% of the particles collected over land contained silver, as compared to less than 5% over the ocean and less than 5% sampled by aircraft. More recent work by Warburton et al. (1975) suggests that the surface layer of the ocean, where substantial elemental enrichment can occur, might supply a background silver concentration in the Antarctic snowpack of between 0.34 and 1.59×10^{-12} g/ml.

As outlined throughout this publication, anthropogenic silver aerosols may be produced by a number of different processes; with fossil fuel combustion, photographic processing, and metal smelting being only a few. The transport processes for background and anthropogenic aerosols are the same. Those particles with appreciable terminal fall velocities (greater than 5 cm/sec) will fall out rapidly. The very small particles with high Brownian diffusion rates will coagulate to form larger aerosols, and the total aerosol spectrum will be subject to losses to the surface from precipitation process interactions such as rainout, washout, scavenging by ice crystals, and particle loss due to activation. During the activation process, aerosols can act as cloud droplet condensation or ice crystal nuclei.

As already indicated, there are numerous industries and processes which use silver directly and/or are capable of dispersing quantities of silver inadvertently. These inadvertent anthropogenic (IA) activities can substantially increase the local silver aerosol concentrations.

Aerosol silver, whether natural or man-made, will eventually leave the atmosphere through dry deposition, washout, rainout, or snowout processes. Limited data prevents formulation of realistic estimates of the magnitude of the dry deposition of silver. Data do exist to make some estimates of the amount deposited by precipitation processes, as summarized in Table 2.2, along with the type and number of precipitation samples, locations, and analytical technique used.

It can be seen that concentrations vary greatly, from <0.1 to 1200×10^{-12} g/ml. Average concentrations reduce variability somewhat, but the concentration range is still wide. This may be partially attributed to sample procurement, handling and analytical procedures, as well as to site location and meteorological factors.

2.4 BACKGROUND ICE NUCLEI

The measurement techniques for determining ice nuclei concentrations yield a variety of results depending on the specific techniques used. Workshops set up to compare a number of different instruments have shown average concentrations measured to range over two to three orders of magnitude (one order of magnitude is equivalent to a factor of 10). It can be seen from these efforts that an absolute measure of ice nuclei concentrations is not possible at this time, however, the currently accepted estimates of background ice nuclei concentrations lie between 10 and 0.1/liter (Kline, 1963; Fletcher, 1969; Chappell, 1970; Bigg, 1973; Allee, 1974). The background ice nuclei detected appear to be of terrestrial origin (Kline, 1963), and a sizeable fraction appear to be of continental origin (Allee, 1974).

2.5 WEATHER MODIFICATION AS A SILVER SOURCE

The concentration of silver from seeding operations in the atmosphere at any given time is a function of the generation characteristics, the source strength, the distance from the source, and the atmospheric conditions (stability, cloud size, cloud precipitation rate, strength of local wind systems, and the past history of the air parcel).

TABLE 2-2. *Background concentrations of silver in precipitation*

Author	Location	Precipitation Form	Number of Samples	Mean Conc. (pg/ml)	Measured
1. Bollay (1965)	Western U.S.	Snow	70	17.4	1-300
2. Schaefer and Fuquay (1965)	Yellowstone Park, Wyoming	Snow	6	40	26-56
3. Engelmann et al. (1967)	Rimrock Lake, Washington	Snow	7	34.5	4.7-74
4. Warburton and Young (1968a,b)	Rapid City, South Dakota	Rain and Hail	7	183*	12-710
	Reno region of Sierra Nevada	Snow	16	All values below detection limit	<8.2-38
5. Parungo and Robertson (1969)	Park Range, Colorado	Snow	19	125	0-1,200
6. Steffe (1970)	South Dakota and Nebraska	Rain and Snow	25		7-150
7. Warburton (1971)	Reno-Lake Tahoe, Nevada	Snow	4	4.5	3.8-5.5
8. Warburton and Young (1972)	Sierra Nevada	Snow	109	4	
	Eastern Nevada	Snow	25	22	
	Bridge Range Mtn.	Snow	27	36	
	Rocky Mtns., Colorado	Snow	48	43	
9. Woodriff et al. (1973)	Bridger Range, Montana	Snow	2	100	60-140
10. Warburton (1973)	Alberta, Canada	Rain and Hail		5	
11. Bogen (1974)	Heidelberg, Germany	Rain	3	3,200	70-9,100
12. Rattonetti (1974)		Rain		50	
13. Gatz (1975)	Southern Illinois	Rain and Snow	82	93*	<2-520
14. Warburton et al. (1975)	Antarctica	Snow	65	0.8	0.14-2.17
15. Teller et al. (1976)	Wolf Creek Pass, Colorado	Snow	49	11	
16. Struempler (1976)	Chadron, Nebraska	Snow	10	48	
	Chadron, Nebraska	Rain	17	90	
17. Mulvey and Sheaffer (1978)	Eastern Plains of Colorado	Snow	142	14.6	<0.1-128

*Average includes values below detection limit that were taken at the detection limit.

1. Neutron Activation
2. Neutron Activation
3. Neutron Activation
4. Flame AA
5. Neutron Activation
6. Flame AA
7. Neutron Activation
8. Neutron Activation
9. Furnace AA
10. Neutron Activation
11. Neutron Activation
12. Flameless AA
13. Flameless AA
14. Neutron Activation and Flameless AA
15. AA
16. AA
17. Flameless AA

2.5.1 Generator Characteristics

A wide range of silver iodide generation systems, both ground-based and airborne, are presently in use. The ground-based generators include electric arc, solid burners (coke impregnated), flares, and solution burners using AgI•NaI, AgI•KI, or AgI•NH_4I with acetone. The aerosols produced by combustion of the AgI•NaI and AgI•KI acetone systems are complexes of silver iodide and the respective alkali iodides possibly in the same molecular ratios present in the acetone solutions. The aerosol from the AgI•NH_4I-acetone system is a relatively simple silver iodide complex. Airborne seeding generators include rockets, flares, and solution burners. Each generation system produces an aerosol with slightly different physical, chemical, and nucleating characteristics, but it is beyond the scope of this work to describe the specific characteristics of each of the above mentioned generator types. It has been shown, however, that the properties of aerosols from generators depend on the conditions of generation, such as ventilation velocity, solution mixture, etc. (Garvey, 1975; Gerber and Allee, 1972; Mossop and Tuck-Lee, 1968; Fletcher, 1969; Parungo et al., 1974; Langer et al., 1974).

The size distribution of the aerosols produced by solution generators is roughly of the form:

$$\frac{dN}{d(\log r)} = C_A r^{-\beta A}, \text{ or of the form } N = D_A r^2 e^{-EAr}$$

The mean particle size varies with burner type, but it generally lies between 0.002 and 0.7 μm. Very hot flames invariably lead to the production of very small particles through silver iodide vaporization and recondensation.

Some types of pyrotechnic flares even appear to have a bimodal distribution, with the smallest particle (0.05 μm diameter) being most numerous and the larger particles (0.2 μm diameter) being a factor of 10 or so less numerous. These size distributions suggest that gravitational settling losses are minimal, and that losses via diffusion to surfaces should predominate outside of cloudy or precipitating zones.

2.5.2 Source Strength

The effective source strength of silver from the generating system will depend on the generator system itself, and on the specific weather modification goal and its associated seeding strategy. Some systems (i.e., flares, rockets, artillery shells) produce large quantities of seeding material over

TABLE 2-3. *Typical ranges of AgI production rates*

Generator Type	Production Rate (g/min)
Electric Arc	0.008-0.04
Solution Burners (ground-based)	0.3-3.0
Solution Burners (airborne)	2.0-20.0
Pyrotechnic Flares	2.0-170.0
Pyrotechnic Rockets	200.0-7,000.0

relatively short periods of time, while others, such as an electric arc system, are usually operated at a much lower production rate.

Typical silver iodide production rates for various types of generators are given in Table 2-3. Artillery shells, used in weather modification programs in the USSR and some other countries, usually contain between 20-500 g of silver iodide per shell.

2.5.3 Distance

The distance from the release point and the physical surroundings of the release point are, as one would expect, two of the major controls on the concentration of seeding material. This is especially true with surface releases in mountainous terrain, where the horizontal gradients of seeding material concentration are often extreme. The principle reason for the strength of distance and surroundings controls on concentration is the dispersion capability of the atmosphere, which is based on atmospheric conditions such as stability, presence and structure of cloud systems, and strength of local wind systems.

Most of the measurements made to track the seeding material through the atmosphere use active ice nuclei concentrations as well as some relative concentrations as indicators of the presence or absence of seeding material. Measurements of the entry and movement of seeding material in the snowpack, soil, or aquatic environment have been confined almost entirely to measurement of silver, with higher than background concentrations being attributed to seeding. Thus, any discussion of the movement of silver iodide using previous work must be based on assumptions of how the measurements actually correspond to silver iodide amounts. Experimental work by Gerber et al. (1970) and Gerber (1974) has shown that only a fraction of the silver iodide generated is detected as an ice nucleant in ice nuclei counters. Therefore, a unique equivalence between the observed ice nuclei concentration and the mass concentration of silver iodide probably does not exist. At best, the ice nuclei concentration may be used only as a relative measure to locate and estimate the structure of seeding material plumes. The measurements of silver in the air, snowpack, soil, or aquatic environment must also be treated as a variable quantity on top of an uncertain background. Concentrations of silver above background levels must be viewed warily, as they may be due to local silver sources. References to above background ice nuclei concentrations and above background silver concentrations throughout this chapter will be equated with silver iodide seeding material. The equating of above background ice nuclei and silver concentrations with seeding material does cause some problems, however, since background levels of this material have not been established.

2.5.4 Atmospheric Conditions

Atmospheric stability is very important in controlling the concentration of seeding material. Inversions are a class of a dispersion-limiting phenomenon, which, whether low level or mid-tropospheric, will act to reduce the verticle dispersion of seeding materials released below them. At the other end of the scale, convection, whether dry or moist, will act to enhance the vertical dispersion of seeding material and reduce its concentration.

Another atmospheric condition which acts to control the concentration of seeding material is cloudiness, estimated to be an important atmospheric aerosol-removing mechanism. This is even more evident when one considers a subset of the aerosol population which acts as a center for nucleation. The transport of seeding material through a cloud system can reduce its concentration considerably, depending on the nucleating effectiveness and physical characteristics of the AgI, the cloud temperature, the residence time of the AgI in the cloud, and the dynamical and microphysical properties of the cloud itself. This aspect of the atmospheric transport process requires that the seeding material be either directly injected into the cloud, or that local wind systems be allowed to carry AgI into the cloud.

The strength of the local wind system is of particular importance in considering the concentration of material released at the surface in mountainous terrain. As has been shown by Reid (1976), the surface topography can induce low-level channeling of the seeding material. Under certain meteorological conditions the vertical transport of seeding material can be drastically reduced; the silver iodide may even be transported upwind of the generator sites. Surface generators not located in mountainous terrain will depend, to a great extent, on meteorological conditions (wind speed, mechanical and thermal turbulence, and atmospheric stability). Accordingly, silver aerosol concentrations and depositional gradients will not be as severe as in the mountainous environment.

The history of the air parcel between the release and observation times is also a controlling factor governing silver concentration due to weather modification activities. Since generators are always operated for a finite period, the timing of the observation becomes critical. This is illustrated by considering the temporal variability of silver immediately after the airborne release of a seeding flare. The flare will burn for a certain time period, emitting silver particles which will be advected with the mean wind and subjected to a spectrum of turbulent eddies, all of which disperse the silver. The result could be an extremely high concentration just downwind of the source a few seconds after ignition. Yet, within several minutes the silver may have passed the observation point and the concentration may again return to ambient levels.

2.6 DEPOSITION

As one might anticipate, the deposition of silver from seeding generators is a function of silver concentration, particle size, state of the atmosphere, distance from the release point, characteristics of the surface, and the precipitation rate.

In treating the key problem of where the seeding material will be deposited, it is convenient to define three regions wherein silver iodide can be deposited on the surface:

1. Immediately surrounding the generator, from the site to a 500 m distance,
2. a local zone including the target site, from 500 m to a 50 km distance, and
3. more distant zones (>50 km).

It has been generally observed (L. O. Grant, personal communication) that the snow surface in the immediate vicinity of surface generators can be yellow in color due to silver iodide. Measurements of silver concentrations in the soil and vegetation within 50 m of generator sites (Teller et al., 1976) have shown significant deposition rates, as shown in Figure 2-2. The rapid falloff of silver concentration with increasing distance from the generator site is very apparent. This rapid, local deposition of silver was probably due to atmospheric turbulence.

The silver concentration in precipitation collected within the local zone varies widely depending on the type of generator used, the collection procedure, source strength, and environmental conditions. For airborne wintertime orographic cloud seeding operations using pyrotechnic flares, Warburton (1971) reported silver concentrations in snow (surface collection) of from 123 to 621 $\times 10^{-12}$ g/ml. These values were between 3 and 30

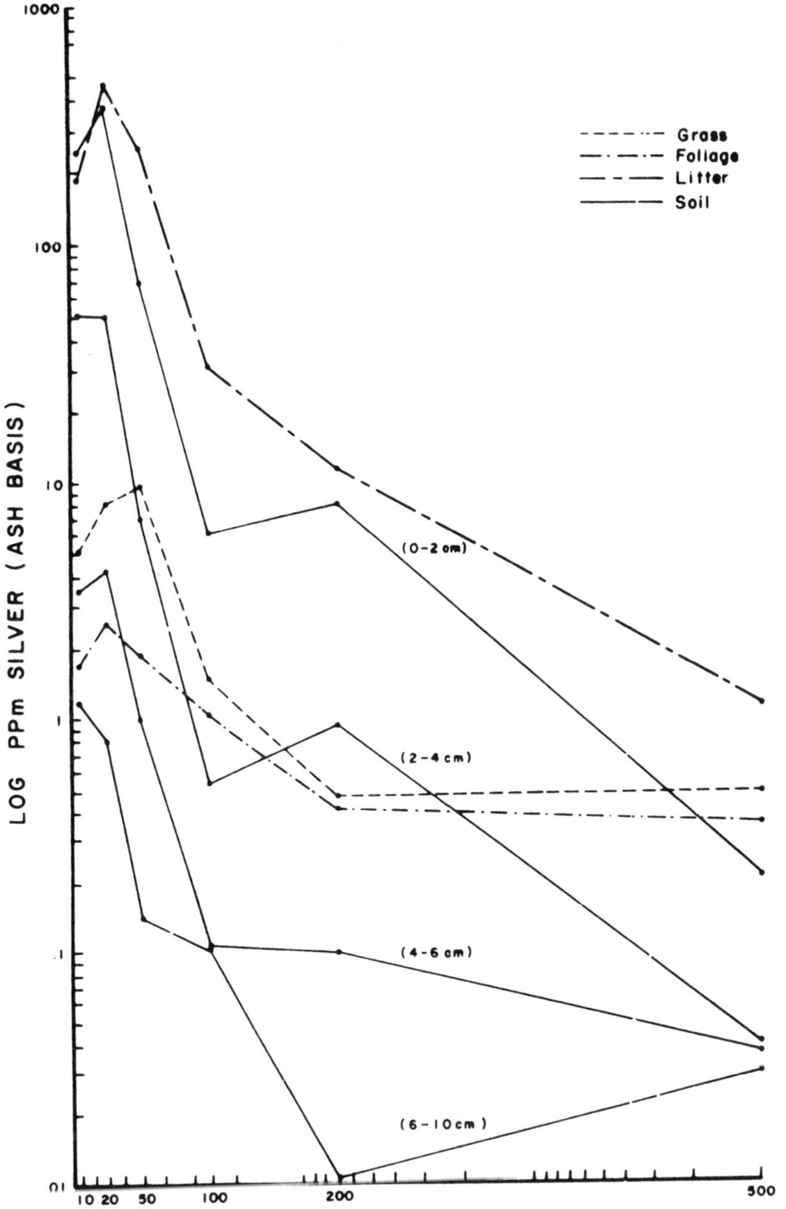

FIGURE 2-2. *Soil silver concentrations as a function of distance from a ground-based generator (Teller et al., 1976).*

times higher than nominal background levels. For summertime airborne cumulus cloud seeding operations using pyrotechnic flares, Summers (1972) reported silver concentrations in rain (surface collection) of from 3 to 42 $\times 10^{-10}$ g/ml. No background could be established due to the instrumental detection limit (1 to 2 $\times 10^{-10}$ g/ml), and all nonseeded events were below detection limits. The results of several studies of the silver content in precipitation are summarized in Table 2-4.

The amount of silver deposited in long distance zones has not been studied extensively. Warburton (1971) was able to detect excess silver in precipitation samples taken over 240 km downwind of seeding operations. The excess silver ranged from 55 to 500% above nominal background values of 6.2 to 24.0 $\times 10^{-12}$ g/ml in the 80 km distant data set, and 123% above background (8.5 $\times 10^{-12}$ g/ml) in the 240 km distant data set. The long transport distances suggested in his work were supported by observations of excess ice nuclei concentrations 400 km downwind of a seeding project (Reinking and Grant, 1968). Observations made by Mulvey and Sheaffer (1978) have also provided evidence of the deposition in snow of seeding material which was transported long distances (100-200 km). They found concentrations on seeded days to average 21.8 $\times 10^{-12}$ g/ml, which was 90% above the average concentration on nonseeded days.

Persistence of silver in rainwater on days following seeding activities does not appear to be a problem. Rainwater sampling on days immediately after seeding has shown that silver levels rapidly return to near-background concentrations (Wisniewski, 1976).

2.7 ATMOSPHERIC AND TOPOGRAPHIC EFFECTS

The transport and dispersion of seeding material is sensitive to the generator location and the state of the atmosphere. Surface-based generators rely on atmospheric currents to carry the seeding material to the target cloud base, while airborne generators usually allow direct injection of seeding materials into the target cloud. The movement of surface-generated silver iodide is largely restricted by the local environmental and topographic conditions. As shown by Reid (1976), low level channeling of seeding material in mountainous terrain by the mountain valley structure is an integral part of the local transport. The importance of local mountain valley breezes on low level inversions in governing dispersions was also demonstrated. In one case, a low level transport of seeding material 40 km upwind of the generator set was accomplished by a low level thermally-driven breeze. Low level inversions were also shown to restrict the movement of seeding material.

Once above the mountain tops, however, the topography acts with the atmospheric stability to control the dispersion of the seeding material. The

TABLE 2-4. *Silver concentrations in precipitation on seeded days*

Reference	Location	Precipitation Form	Site	# of Samples	Ag CONC x 10^{-12} g/ml Mean	Range
Parungo and Robertson (1969)	Park Range, Colorado	Snow	Within Target	73	9,500	0 - 80,000
Warburton and Young (1968)	South Dakota	Rain and Hail	Within Target	4	N/A	25 - 220
Summers (1972)	Alberta, Canada	Rain and Hail	Within Target		N/A	<200 - 3,800
		Rain and Hail	Intermediate Downwind		N/A	<200 - 1,400
		Rain and Hail	Far Downwind		N/A	<200 - 4,200
Gatz (1975)	Southern, Illinois	Rain and Snow	Within Target	82	93	<2 - 520
Warburton (1971)	Steamboat Springs, Colorado	Snow	Within Target		217	
	Lake Erie, New York	Snow	Within Target		621	
	Reno-Lake Tahoe Nevada	Snow	Within Target		159	
	Reno-Lake Tahoe Nevada	Snow	Intermediate Downwind		8.7	
Mulvey and Sheaffer (1978)	Eastern Plains and Foothills of Colorado	Snow	Far Distant	62	21.8	< 0.1 - 128

short-range plume tracking experiments of Langer et al. (1967), carried out under stable conditions, show more modest horizontal dispersion rates but more considerable vertical dispersion. On the other hand, long-range seeding transport as observed by Mulvey and Grant (1976) shows that extensive horizontal dispersion can occur under stable conditions.

2.8 DEPOSITION HETEROGENEITY

As has been pointed out, the deposition of seeding material is far from uniform. High concentrations of material can be deposited within 500 m of the generator site (for ground-based generators) in the absence of active convection; outside of this area lesser amounts of seeding material are deposited. However, local wind field and/or topographic features controlling the dispersion can cause considerable variation in horizontal deposition within the primary target area. Outside of the primary target area the deposition is quite spotty, since the dispersion rate and transport direction changes make the concentration of seeding material quite variable.

Also, a higher variability in the deposition of silver should occur in summer than in wintertime programs. The differences between summer and winter programs are not only due to differences in the general state of the atmosphere and predominant cloud types, but also in the rationale, method, and quantity of material used for seeding. In general, one would expect seeding material to be deposited in wider areas in summer due to the more transient nature of the target clouds.

2.9 TIMING

Several of the factors affecting the time of entry of silver into the snowpack or aquatic environment will be discussed in other chapters. Atmospheric conditions, depending on the seeding strategy, may have a major influence on the time of entry. Ground generator effluent can be isolated from target clouds by low level inversions and local winds, where most of the transport to the surface occurs. Airborne operations avoid the inversion problem, but unless the AgI is directly injected into the target cloud, local winds can delay the interaction.

As noted previously, the aerosol size essentially rules out gravitational settling as a downward transport mechanism. Small particles which do not enter clouds will practically remain suspended indefinitely; silver iodide has been observed to remain suspended in noncloud target areas for considerable time periods (Rottner et al., 1975; Reid, 1976). Even particles which do not enter clouds are not necessarily transported to the surface. If precipitable particles evaporate or completely sublime, the AgI particles can be reintroduced into the atmosphere to await other cloud systems. Seeding material which does reach a precipitating cloud will normally be transported to the surface within several hours.

2.10 CONCLUSIONS

The material presented in this chapter indicates that the background silver concentrations in air and in precipitation are highly variable. The specific weather modification hypothesis and its associated seeding technology (generator type, seeding rate, and generator location) play a major role in determining silver concentrations in air and in precipitation due to weather modification activities. The resulting combination of a highly variable background level and a wide range of seeding technologies presently in use makes the detection of seeding material based on silver concentrations extremely difficult. Such detection techniques require a long period of nonseeded background measurements as well as seeding measurements to identify seeding material in potential impact areas.

Standard collection, preparation, and analytical procedures will be helpful to allow better measurement of silver concentrations in air and precipitation. The release of tracer materials, to differentiate silver from seeding activities from natural and other anthropogenic sources, is also highly recommended. Problems related to the movement and deposition of seeding materials, particularly in mountainous terrain, should continue to receive research emphasis.

Chapter 3

Theoretical Solubility Relationships of Silver in Soils

W. L. Lindsay and M. Sadiq

3.1 INTRODUCTION

The use of silver iodide as a nucleating agent for cloud seeding has raised many questions regarding the fate of silver (Ag) in the environment (Weaver and Klarich, 1973; Cameron, 1973; Jones and Bailey, 1974; Molise and Klein, 1974; Roy and Bailey, 1974). A descriptive review of silver in the environment has been given by Smith and Carson (1977).

Some of the leading questions are:

1. What silver minerals are formed in soils?
2. Are these minerals sufficiently soluble to be taken up by plants or moved into underground waters?
3. Will future changes in oxidation-reduction status, salinity, and pH adversely affect the mobility and availability of silver?

The purpose of this chapter is to summarize the present knowledge of silver into meaningful solubility diagrams and equations, in order to predict probable and improbable forms of silver in soils. Hopefully, this information will aid in understanding silver equilibria in soils and other aqueous environments.

3.2 SELECTION OF ΔG_f° VALUES FOR SILVER SPECIES

The parameter often used to characterize the solubility relationships of chemical elements is the standard free energy of formation (ΔG_f°) of the various chemical species. An exhaustive search of literature was made to obtain the best ΔG_f° values for silver minerals and aqueous species, as part of a more comprehensive attempt to select ΔG_f° values for use in soil science

(Sadiq and Lindsay, 1978). The selected values for silver and its associated species are summarized in Table 3-1.

3.3 CALCULATION OF EQUILIBRIUM CONSTANTS FOR SILVER REACTIONS

The equilibrium constants of chemical reactions involving silver minerals and its aqueous solution species are summarized in Table 3-2. These constants were calculated from the ΔG_f^o values given in Table 3-1 using the relationship:

$$\log K^o = -\frac{\Delta G_r^o}{1.364} \qquad (1)$$

where log K° is the equilibrium constant at zero ionic strength, ΔG_r^o = $\Sigma\Delta G_f^o$ of products minus $\Sigma\Delta G_f^o$ of reactants in kcal mole^{-1} for the reaction being considered. The constant 1.364 relates to 25°C and other thermodynamic parameters. Further details on the calculation of equilibrium constants and their use in soils are given by Lindsay (1978). A sample calculation of the log K° of Reaction 5 in Table 3-2 is:

$$Ag^+ + H_2O \rightleftharpoons AgOH^o + H^+$$

$$\Delta G_r^o = (-21.84) + (0) - (18.465) - (-56.687)$$

$$= 16.382$$

$$\log K^o = -\frac{\Delta G_r^o}{1.364} = -\frac{16.382}{1.364} = -12.01$$

Thus, the log K° for this reaction is –12.01. The other log K° values in Table 3-2 were calculated similarly. Equilibrium constants provide the essential information for determining the solubility relationships of silver minerals and solution species in soils.

3.4 EFFECT OF REDOX ON THE SOLUBILITY OF SILVER MINERALS

Silver exists in four oxidation states: 0, +1, +2, and +3. The oxidation states 0 and +1 are of greatest importance in soils as shown in Figure 3-1 (constructed by combining appropriate equilibrium reactions from Table 3-2).

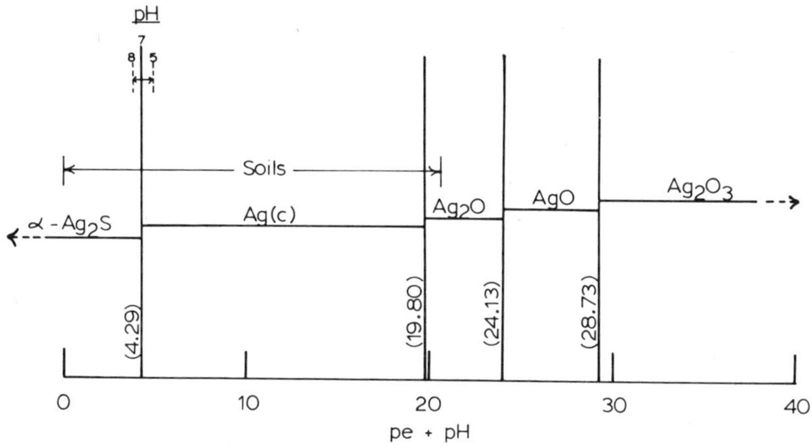

FIGURE 3-1. *The redox stability range of various silver minerals.*

The redox parameter used in Figure 3-1 is pe + pH, where pe is negative log of the electron activity, and pH is the negative log of proton activity (Lindsay and Sadiq, 1978). The redox range labeled "soils" in this figure corresponds to the stability region of water; where pe + pH is equal to 0 in equilibrium with 1 atm $H_2(g)$ and 20.78 in equilibrium with 1 atm of $0_2(g)$. Of the minerals shown in Figure 3-1, metallic Ag(c) is most stable in the redox range of soils and most natural aqueous environments. Only in highly reduced soils would α-Ag_2S be expected to form. The transition redox between Ag(c) and α-Ag_2S shifts slightly, depending upon the activity of SO_4^{2-} and pH. An SO_4^{2-} activity of 10^{-3} M and a pH of 7 give a transition pe + pH of 4.29 for this transformation as shown in Figure 3-1. The α-Ag_2S is only slightly more stable than β-Ag_2S (Reactions 41 and 42 of Table 3-2). Both minerals may occur in highly reduced environments due to their similar solubilities.

Only under highly oxidized environments, not normally encountered in soils, would silver oxides be stable (Figure 3-1): Ag_2O is stable from pe + pH 19.80 to 24.13, AgO from 24.13 to 28.73, and Ag_2O_3 above pe + pH 28.73. None of these oxides are expected to persist in soils.

3.5 EFFECT OF REDOX ON SILVER IONS

The predominant silver ion expected in soils and aqueous environments is Ag^+. From Reaction 2 of Table 3-2:

$$Ag^+ \rightleftharpoons Ag^{2+} + e^- \qquad (\log K^\circ = -33.60) \qquad (2)$$

TABLE 3-1. *Selected standard free energies of formation of silver species.*

Species	ΔG_f^o	Source
	kcal $mole^{-1}$	
	Solution Species	
Ag^+	18.465	Anderson et al. (1971)
Ag^{2+}	64.3	Wagman et al. (1969)
$AgBr^o$	- 12.38	Berne and Leden (1953)
$AgBr_2^-$	- 41.29	Berne and Leden (1953)
$AgBr_3^{2-}$	- 66.99	Berne and Leden (1953)
$AgBr_4^{3-}$	- 92.88	Berne and Leden (1953)
$AgCl^o$	- 17.42	Jonte and Martin (1952)
$AgCl_2^-$	- 51.44	Jonte and Martin (1952)
$AgCl_3^{2-}$	- 82.58	Berne and Leden (1953)
$AgCl_4^{3-}$	-114.27	Berne and Leden (1953)
AgF^o	- 48.36	Connick and Paul (1961)
AgI^o	- 3.07	Wagman et al. (1969)
AgI_2^-	- 21.17	Wagman et al. (1969)
AgI_3^{2-}	- 37.4	Wagman et al. (1969)
AgI_4^{3-}	- 50.87	Wagman et al. (1969)
$AgNH_3^+$	7.52	Vosburgh and McClure (1943)
$Ag(NH_3)_2^+$	- 4.07	Vosburgh and McClure (1943)
$AgNO_3^o$	- 7.76	Wagman et al. (1969)
$AgOH^o$	- 21.84	Biedermann and Sillen (1960)
$Ag(OH)_2^-$	- 62.20	Biedermann and Sillen (1960)
$Ag(OH)_3^{2-}$	- 85.91	Dirkse and Graven (1959)
$AgHS^o$	2.14	Cloke (1963)
$Ag(HS)_2^-$	- 3.03	Cloke (1963)
$AgSO_3^-$	-105.17	Wagman et al. (1969)
$AgSO_4^-$	-161.28	Hopkins and Wulff (1965)
$Ag_2SO_3^o$	- 91.21	Chateau et al. (1956a)
$Ag_2SO_4^o$	-141.04	Wagman et al. (1969)
$Ag_3(OH)_4^-$	-120.78	Pleskov and Kabanov (1957)
	Minerals	
Ag(c)	0.00	CODATA (1976)
AgBr(c)	- 23.18	Ramette (1972)
AgCl(c)	- 26.244	Pankratz (1970)
AgF(c)	- 44.39	Peacock (1959)
AgI(c)	- 16.04	McIntyre and Amis (1968)

(continued)

TABLE 3-1. *(continued)*

Species	ΔG_f^o	Source
	Minerals (continued)	
$AgNO_3(c)$	- 7.98	Parker (1965)
		Latimer et al. (1938)
$AgO(c)$	3.23	Dirkse (1962)
$AgOH(c)$	- 29.65	Antikainen and Dryssen (1960)
$Ag(OH)_2(c)$	- 45.77	Dirkse and Graven (1959)
$Ag_2CO_3(c)$	-104.36	Kelley and Anderson (1935)
$Ag_2O(c)$	- 2.68	Gregor and Pitzer (1962)
$Ag_2O_2(c)$	6.6	Wagman et al. (1969)
$Ag_2O_3(c)$	28.14	Stehlik (1963)
α-$Ag_2S(c)$	- 9.51	Mustafaev and Abbasov (1970)
β-$Ag_2S(c)$	- 9.45	Rosenquist (1949)
$Ag_2SO_3(c)$	- 98.22	Chateau et al. (1956b)
$Ag_2SO_4(c)$	-147.60	Lietzke and Hall (1967)
$Ag_3PO_4(c)$	-211.57	Wagman et al. (1969)
	Other Species	
Br^-	- 24.87	CODATA (1971)
Cl^-	- 31.371	CODATA (1971)
$CO_2(g)$	- 94.26	Gordon (1961)
CO_3^{2-}	-126.16	Adams (1971)
		Garrels et al. (1961)
e^-	0	Standard state
F^-	- 67.34	CODATA (1976)
H^+	0	Standard state
$H_2O(l)$	- 56.687	Wagman et al. (1968)
I^-	- 12.38	CODATA (1971) and Johnson (1975)
NH_4^+	- 18.97	Wagman et al. (1968)
NO_3^-	- 26.61	Wagman et al. (1968)
PO_4^{3-}	-245.17	Vanderzee and Quist (1961)
S^{2-}	20.5	Garrels and Naeser (1958)
HS^-	2.88	Kury et al. (1953)
SO_3^{2-}	-116.30	Cuta et al. (1957)
SO_4^{2-}	-177.97	Roth (1932)
		Wagman et al. (1968)

TABLE 3-2. *Equilibrium reactions of Ag minerals and solution species in aqueous environments calculated from the ΔG_f^o values in Table 3-1*

Reaction No.	Chemical Reaction		log K°
	Redox Reactions		
1	$Ag(c)$	$\rightleftharpoons Ag^+ + e^-$	-13.54
2	Ag^+	$\rightleftharpoons Ag^{2+} + e^-$	-33.60
	Ag(I) Oxides and Hydrolysis		
3	$Ag_2O(c) + 2H^+$	$\rightleftharpoons 2Ag^+ + H_2O$	12.52
4	$AgOH(c) + H^+$	$\rightleftharpoons Ag^+ + H_2O$	6.28
5	$Ag^+ + H_2O$	$\rightleftharpoons AgOH^o + H^+$	-12.01
6	$Ag^+ + 2H_2O$	$\rightleftharpoons Ag(OH)_2^- + 2H^+$	-23.98
7	$Ag^+ + 3H_2O$	$\rightleftharpoons Ag(OH)_3^{2-} + 3H^+$	-48.16
8	$3Ag^+ + 4H_2O$	$\rightleftharpoons Ag_3(OH)_4^- + 4H^+$	-37.08
	Ag(II) Oxides		
9	$AgO(c) + 2H^+$	$\rightleftharpoons Ag^{2+} + H_2O$	- 3.21
10	$Ag(OH)_2(c) + 2H^+$	$\rightleftharpoons Ag^{2+} + 2H_2O$	2.42
	Ag(III) Oxides		
11	$1/2Ag_2O_3(c) + H^+ + e^-$	$\rightleftharpoons AgO(c) + 1/2H_2O$	28.73
	Bromides		
12	$AgBr(c)$	$\rightleftharpoons Ag^+ + Br^-$	-12.30
13	$Ag^+ + Br^-$	$\rightleftharpoons AgBr^o$	4.38
14	$Ag^+ + 2Br^-$	$\rightleftharpoons AgBr_2^-$	7.34
15	$Ag^+ + 3Br^-$	$\rightleftharpoons AgBr_3^{2-}$	7.95
16	$Ag^+ + 4Br^-$	$\rightleftharpoons AgBr_4^{3-}$	8.70
	Chlorides		
17	$AgCl(c)$	$\rightleftharpoons Ag^+ + Cl^-$	- 9.78
18	$Ag^+ + Cl^-$	$\rightleftharpoons AgCl^o$	3.31
19	$Ag^+ + 2Cl^-$	$\rightleftharpoons AgCl_2^-$	5.25
20	$Ag^+ + 3Cl^-$	$\rightleftharpoons AgCl_3^{2-}$	5.08
21	$Ag^+ + 4Cl^-$	$\rightleftharpoons AgCl_4^{3-}$	5.32
	Fluorides		
22	$AgF(c)$	$\rightleftharpoons Ag^+ + F^-$	3.29
23	$Ag^+ + F^-$	$\rightleftharpoons AgF^o$	- 0.38

(continued)

TABLE 3-2. *(continued)*

Reaction No.	Chemical Reaction		log K°
	Iodides		
24	$AgI(c)$	$\rightleftharpoons Ag^+ + I^-$	-16.22
25	$Ag^+ + I^-$	$\rightleftharpoons AgI^o$	6.71
26	$Ag^+ + 2I^-$	$\rightleftharpoons AgI_2^-$	10.91
27	$Ag^+ + 3I^-$	$\rightleftharpoons AgI_3^{2-}$	13.73
28	$Ag^+ + 4I^-$	$\rightleftharpoons AgI_4^{3-}$	14.53
	Ammonia Complexes		
29	$Ag^+ + NH_4^+$	$\rightleftharpoons AgNH_3^+ + H^+$	- 5.88
30	$Ag^+ + 2NH_4^+$	$\rightleftharpoons Ag(NH_3)_2^+ + 2H^+$	-11.29
	Nitrates		
31	$AgNO_3(c)$	$\rightleftharpoons Ag^+ + NO_3^-$	0.12
32	$Ag^+ + NO_3^-$	$\rightleftharpoons AgNO_3^0$	-0.28
	Carbonates		
33	$Ag_2CO_3(c) + 2H^+$	$\rightleftharpoons 2Ag^+ + CO_2(g) + H_2O$	7.08
	Phosphates		
34	$Ag_3PO_4(c)$	$\rightleftharpoons 3Ag^+ + PO_4^{3-}$	-15.98
	Sulfates		
35	$Ag_2SO_4(c)$	$\rightleftharpoons 2Ag^+ + SO_4^{2-}$	- 4.81
36	$Ag^+ + SO_4^{2-}$	$\rightleftharpoons AgSO_4^-$	1.30
37	$2Ag^+ + SO_4^{2-}$	$\rightleftharpoons Ag_2SO_4^0$	0.00
	Sulfites		
38	$Ag_2SO_3(c)$	$\rightleftharpoons 2Ag^+ + SO_3^{2-}$	-13.82
39	$Ag^+ + SO_3^{2-}$	$\rightleftharpoons AgSO_3^-$	5.38
40	$2Ag^+ + SO_3^{2-}$	$\rightleftharpoons Ag_2SO_3^0$	8.68
	Sulfides		
41	$\alpha\text{-}Ag_2S(c)$	$\rightleftharpoons 2Ag^+ + S^{2-}$	-49.08
42	$\beta\text{-}Ag_2S(c)$	$\rightleftharpoons 2Ag^+ + S^{2-}$	-49.03
43	$Ag^+ + HS^-$	$\rightleftharpoons AgHS^o$	14.08
44	$Ag^+ + 2HS^-$	$\rightleftharpoons Ag(HS)_2^-$	19.98
	Other Reactions		
45	$S^{2-} + 4H_2O$	$\rightleftharpoons SO_4^{2-} + 8e^- + 8H^+$	-20.73
46	$SO_4^{2-} + 2e^- + 2H^+$	$\rightleftharpoons SO_3^{2-} + H_2O$	-23.90

This reaction indicates that the ratio of Ag^+/Ag^{2+} is unity at pe = 33.60. In moderately well-oxidized soil where pe is approximately 11, the ratio of Ag^+/Ag^{2+} would be $10^{33.6-11.0} = 10^{22.6}$. The Ag^{3+} would be even less abundant than Ag^{2+}. Thus, the chemistry of Ag in soils and other aqueous environments is largely limited to Ag(c) and Ag^+ with its associated minerals and solution complexes.

3.6 SOLUBILITY OF SILVER HALIDE MINERALS

The Ag^+ ion combines with the halide anions Br^-, Cl^-, and I^- to form relatively insoluble minerals. The F^- salt of silver is considerably more soluble (Reaction 22, Table 3-2). The solubility relationships of Ag halides are shown in Figure 3-2. Superimposed in this diagram is the Ag^+ activity in equilibrium with Ag(c) as affected by the redox potential.

From Figure 3-2 it is easy to predict which silver halides are stable in soils when the halide anion activities and redox are known. The corresponding Ag^+ maintained by each of the silver halides can be readily estimated from the halide anion activities. The halide maintaining the lowest Ag^+ activity becomes the stable phase and others will dissolve; if the redox is sufficiently low that Ag(c) maintains a lower Ag^+ than any of the halides, then Ag(c) becomes the stable phase and all the Ag halides disappear.

As examples, let us consider the four soils depicted in Table 3-3. For Soil A, AgCl maintains the lowest Ag^+ activity ($10^{-8.8}$ M) and would therefore be the stable Ag mineral. In Soil B, AgBr would be the stable Ag mineral maintaining Ag^+ at $10^{-8.8}$ M. In Soil C, AgI would be the stable Ag mineral and maintain Ag^+ at $10^{-10.2}$ M. The redox of Soils A, B, and C is sufficiently high that Ag(c) is unstable in all of them. In Soil D, however, Ag(c) would be the stable Ag mineral maintaining Ag^+ at $10^{-12.5}$ M. None of the halides would be stable in Soil D. Once it is determined that Ag(c) is the stable silver mineral, the activity of Ag^+ can be readily obtained from the relationship:

$$Ag(c) \rightleftharpoons Ag^+ + e^- \qquad (\log K^\circ = -13.54) \tag{3}$$

from which we obtain:

$$\log Ag^+ = -13.54 + pe \tag{4}$$

Another mineral included in Figure 3-2 is α-Ag_2S. The equilibrium between this mineral and Ag(c) can be obtained by combining the reactions:

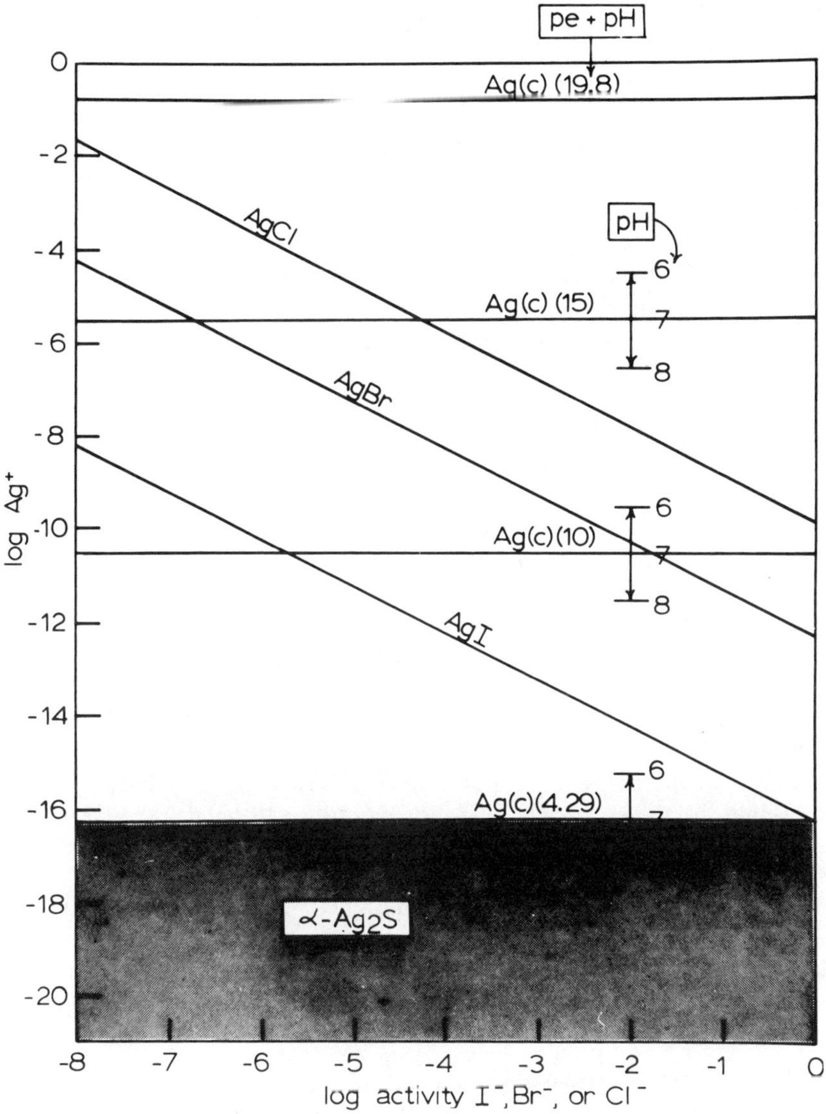

FIGURE 3-2. *The relative stabilities of several silver minerals as affected by halide ion activities, redox, and pH when SO_4^{2-} is 10^{-3} M.*

$\alpha\text{-}Ag_2S(c) \rightleftharpoons 2Ag^+ + S^{2-}$	-49.08
$S^{2-} + 4H_2O \rightleftharpoons SO_4^{2-} + 8e^- + 8H^+$	-20.73
$2Ag^+ + 2e^- \rightleftharpoons 2Ag(c)$	2(13.54)
$\alpha\text{-}Ag_2S(c) + 4H_2O \rightleftharpoons 2Ag(c) + SO_4^{2-} + 6e^- + 8H^+$	-42.73

from which the following equation is derived:

$$pe + pH = 7.12 + 1/6 \log SO_4^{2-} - 1/3\ pH \quad (5)$$

In a soil of pH 7 having 10^{-3} M SO_4^{2-}, the pe + pH at which Ag(c) and α-Ag_2S can coexist is 4.29. This line sets the upper redox limit for the stability of α-Ag_2S as shown in Figure 3-2. Both pH and SO_4^{2-} activity shift this line, as dictated by Equation 5.

Thus, Figure 3-2 gives a convenient graphic means of determining which silver minerals are stable in soils when pH, redox, and halide ion activities are known. As mentioned previously, silver fluoride is much too soluble (Reaction 22 of Table 3-2) to be of importance in soils.

3.7 STABILITY OF OTHER SILVER MINERALS

The solubilities of other silver minerals given by Reactions 31, 32, 34, 35, and 38 of Table 3-2 are plotted in Figure 3-3. In this development, the activity of anionic constituents are arbitrarily set at levels typical of those found in soils. These selections were discussed in detail (Lindsay, 1978).

The solubility of $AgNO_3$ shown in Figure 3-3 corresponds to a NO_3^- range of 10^{-4} to 10^{-2} M. As seen, the Ag^+ activity would have to be in the molar range for $AgNO_3$ to form in soils; it can be discounted as a soil mineral.

The solubility line for Ag_2SO_4 shown in Figure 3-3 was drawn on the basis that the activity of SO_4^{2-} in soils lies between 10^{-4} and 10^{-2} M, respectively, for this mineral to be stable. Thus, Ag_2SO_4 is too soluble to persist in soils and can be discounted as a soil mineral.

The solubility of Ag_2CO_3 shown in Figure 3-3 was drawn on the basis that CO_2(g) ranges from 0.0003 to 0.03 atm, the lower value being the CO_2 level in normal air. At pH 8, the Ag^+ activity in equilibrium with this mineral is $10^{-3.2}$ M and increases at lower pH values. It is obvious that Ag_2CO_3 also is too soluble to persist in soils.

TABLE 3-3. *Sample soils for discussion of Ag minerals and complexes.*

			log activity, mole/l		
Soil	pH	pe + pH	Cl^-	Br^-	I^-
A	6.5	18	-1	-5	<-8
B	7.5	15	-3	-3.5	<-8
C	7.0	18	-3	-5	-6
D	7.0	8	-3	-5	-6

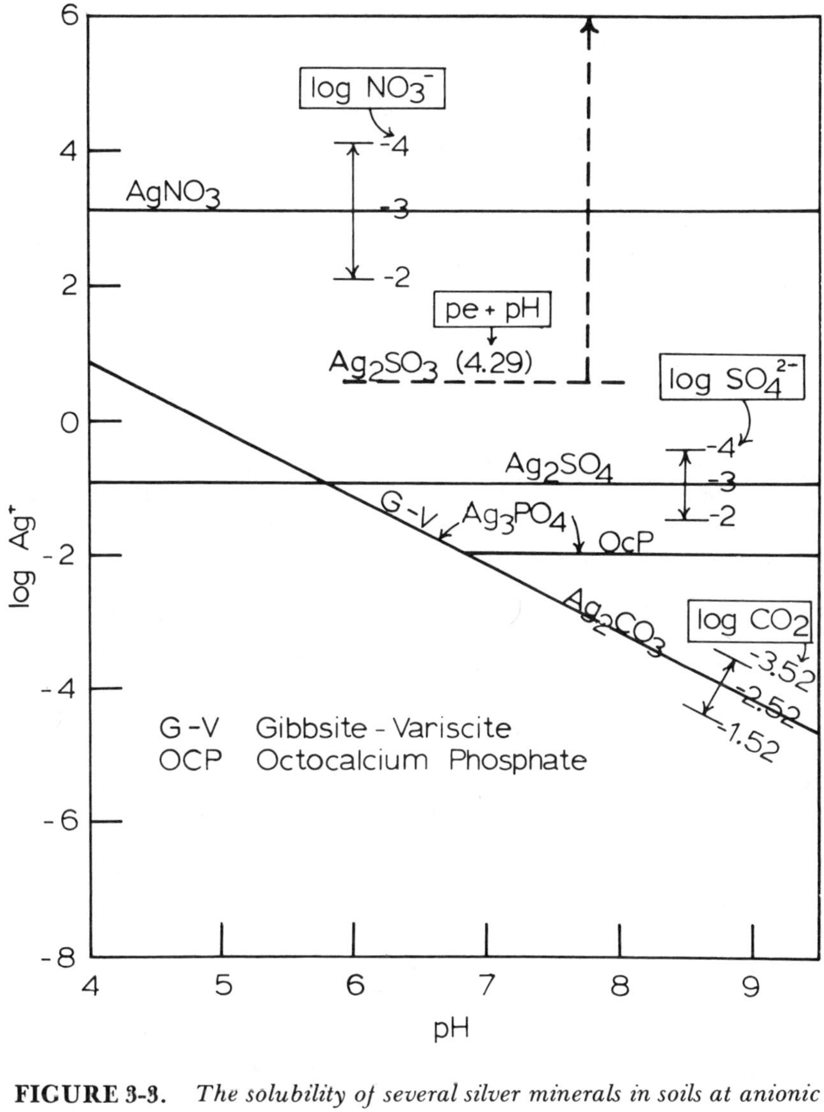

FIGURE 3-3. *The solubility of several silver minerals in soils at anionic activities and redox levels commonly found.*

The solubility of Ag_3PO_4 is also represented in Figure 3-3. At low pH, P activity was assumed to be controlled by $AlPO_4 \cdot 2H_2O$ (variscite) and $Al(OH)_3$ (gibbsite). Thus Ag_3PO_4 would support an Ag^+ activity identical to Ag_2CO_3 at 0.0003 atm CO_2. Above pH 6.8 $Ca_4H(PO_4)_3 \cdot 5/2H_2O$ (octocalcium phosphate) was selected as an appropriate control of phosphate activity. In this case, Ag_3PO_4 controlled Ag^+ at about 10^{-2} M. Again,

Ag_3PO_4 is too soluble to be of importance in soils. Lower phosphate levels would make it even less stable.

For Ag_2SO_3 the equilibrium level of SO_3^{2-} expected in soils range from 10^{-45} M at pe + pH of 20 to 10^{-15} M for pe + pH of 4.29 where Ag_2S becomes stable (Lindsay, 1978). This means that Ag^+ activity would have to be $> 10^{0.59}$ M which is far too high for soils. The formation of Ag_2SO_3 in soils can be discounted.

From Figure 3-3 it may be concluded that none of the minerals shown are important in soils.

3.8 STABILITY OF SILVER HALIDE AND HYDROXYL COMPLEXES

As indicated earlier, Ag^+ is the predominant ionic form of silver in the oxidation range of soils, but how important are the complexes of silver in the soil solution? Figure 3-4 was constructed to show the important halide and hydroxyl complexes of Ag^+.

For this development an arbitrary Ag^+ activity of 10^{-10} M was chosen. Smith and Carson (1977) reported an average silver content for fresh water as 0.2 μg/l which would correspond to a total soluble level of $10^{-8.73}$ M. Since some of this silver is most likely complexed, a selection of 10^{-10} M for Ag^+ activity seems appropriate.

From Figure 3-4 the activities of the various silver halide complexes can be readily determined as a function of the halide ion activities, and the hydroxyl complexes from pH. Only the predominant regions of each complex ion are shown. These lines can be extended to show the activities of the complex ions in concentration regions where they are not the most abundant species. Halide and hydroxyl complexes (Table 3-2), which are not shown in Figure 3-4, require greater than 1 M activities of the associated anions in order to become predominant species. Since such levels are impractical for soils, the graph was not extended to include them.

Even though Figure 3-4 was constructed with Ag^+ at 10^{-10} M, its use is not restricted to this Ag^+ activity. If, for example, Ag^+ activity of a soil were 10^{-8} M, the activities of the complexes need only be multiplied by $10^{-8}/10^{-10}$ or 10^2. Thus, the graph can be used to obtain the activity of silver complexes associated with any Ag^+ activity.

Let us consider the use of Figure 3-4 to estimate the activities of silver complexes in Soil C of Table 3-3. From Figure 3-4, Ag^+ in this soil is estimated at $10^{-10.2}$ M, and AgI° at $10^{-9.5}$ M. Extrapolating AgI_2^- gives $10^{-12.0\ -\ 0.2} = 10^{-12.2}$ M and extrapolating AgI_3^{2-} gives $10^{-14.3\ -\ 0.2} = 10^{-14.5}$ M. The activities of each of the other silver complexes can be similarly estimated. In this soil AgI° comprises most of the soluble silver. The ratio of AgI° to Ag^+ is $10^{-9.5}/10^{-10.2} = 10^{0.7}$ or about five fold.

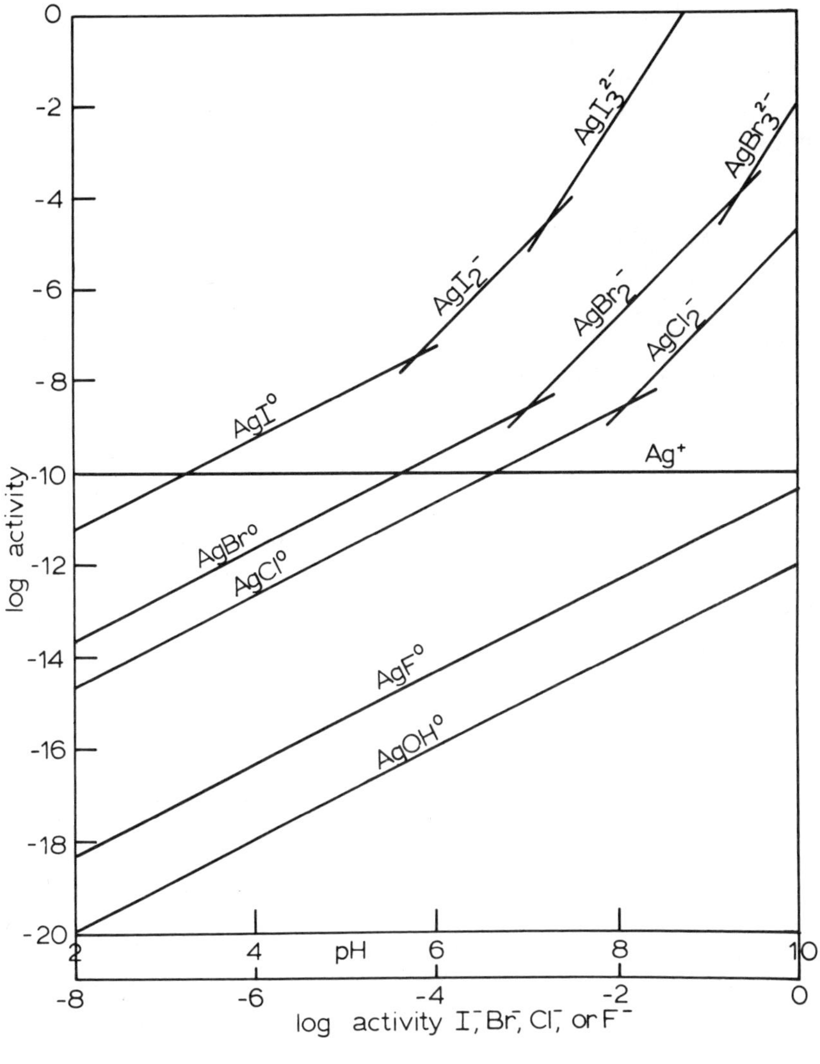

FIGURE 3-4. *The activity of silver complexes in equilibrium with 10^{-10} M silver.*

Thus, Figure 3-2 permits a ready diagnosis of the silver minerals which can exist in soils, and the solubility of Ag^+ that will be maintained. Figure 3-4 can then be used to estimate the activity of the halide and hydroxyl complexes that can be present.

3.9 STABILITY OF OTHER SILVER COMPLEXES

Since Ag^+ forms complexes with several other anions, these complexes will be examined, in the order in which they appear in Table 3-2 to determine their importance in soils.

From Reaction 29 of Table 3-2 comes the relationship:

$$\log AgNH_3^+ = -5.88 + pH + \log NH_4^+ + \log Ag^+ \tag{6}$$

In well-aerated soils the thermodynamic stability of NH_4^+ would be less than 10^{-15} M above pe + pH of 13 (Lindsay, 1978); therefore, the $AgNH_3^+$ complex would not be important under such conditions. Even if NH_4^+ exists metastably in the range of 10^{-3} M as a result of the mineralization of organic matter or application of ammonium fertilizer, then Equation 6 for a soil of pH 7 becomes:

$$\log AgNH_3^+ = -1.88 + \log Ag^+ \tag{7}$$

Under these conditions, the ratio of $AgNH_3^+/Ag^+$ is $10^{-1.88}$. Thus $AgNH_3^+$ becomes slightly significant as modest levels of NH_4^+ occur in soils.

From Reaction 30 of Table 3-2 can be obtained:

$$\log Ag(NH_3)_2^+ = -11.29 + 2\log NH_4^+ + 2pH + \log Ag^+ \tag{8}$$

If NH_4^+ approached 10^{-3} M in soils of pH 7, Equation 8 becomes:

$$\log Ag(NH_3)_2^+ = -3.29 + \log Ag^+ \tag{9}$$

which corresponds to an $Ag(NH_3)_2^+/Ag^+$ ratio of $10^{-3.29}$. The importance of this species can be fairly well discounted in soils.

From Reaction 32 of Table 3-2 comes the equation:

$$\log AgNO_3^{\circ} = -0.28 + \log NO_3^- + \log Ag^+ \tag{10}$$

For a typical NO_3^- level of 10^{-3} M in soils this equation becomes:

$$\log AgNO_3^{\circ} = -3.28 + \log Ag^+ \tag{11}$$

which corresponds to an $AgNO_3^{\circ}/Ag^+$ ratio of $10^{-3.28}$. This complex would not increase the solubility of Ag significantly in soils.

From Reaction 36 of Table 3-2 comes the equation:

$$\log AgSO_4^- = 1.30 + \log SO_4^{2-} + \log Ag^+ \quad (12)$$

For a sulfate level of 10^{-3} M this equation becomes:

$$\log AgSO_4^- = -1.70 + \log Ag^+ \quad (13)$$

which gives a ratio of $AgSO_4^-/Ag^+$ of 1/50, which would increase Ag solubility by only 2%.

For $Ag_2SO_4^o$ complex, Reaction 37 of Table 3-2 gives:

$$\log Ag_2SO_4^o = -3.00 + 2\log Ag^+ \quad (14)$$

For soils where Ag^+ is in the range of 10^{-10} M, the ratio of $Ag_2SO_4^o/Ag^+$ is 10^{-23} so this species is of no importance in soils.

The sulfite complexes of silver shown in Reactions 39 and 40 of Table 3-2 can also be shown to be of no great significance in soils because of the low SO_3^{2-} levels found in soils.

The $AgHS^o$ given by Reaction 43 of Table 3-2 can be combined to give the following equilibria:

		log K°	
$Ag^+ + HS^-$	$\rightleftharpoons AgHS^o$	14.08	
$H^+ + S^{2-}$	$\rightleftharpoons HS^-$	12.92	
$SO_4^{2-} + 8e^- + 8H^+$	$\rightleftharpoons S^{2-} + 4H_2O$	20.73	
$Ag^+ + 9H^+ + 8e^- + SO_4^{2-}$	$\rightleftharpoons AgHS^o + 4H_2O$	47.73	(15)

which gives:

$$\log AgHS^o = 47.73 + \log SO_4^{2-} - pH - 8(pe + pH) + \log Ag^+ \quad (16)$$

For a soil of pH 7 having 10^{-3} M SO_4^{2-} this equation becomes:

$$\log AgHS^o = 37.73 - 8(pe + pH) + \log Ag^+ \quad (17)$$

From this equation the ratio of $AgHS^o/Ag^+$ is unity when $pe + pH = 4.72$ and increases 10^8 for each unit decrease in pe + pH. This means that $AgHS^o$ can be discounted as being important in soils of $pe + pH > 5$, but the complex becomes extremely important below pe + pH of 4.72. It should be recalled that α-Ag_2S precipitates at $pe + pH = 4.29$. At this redox the ratio of $AgHS^o/Ag^+$ is $10^{3.41}$ or 2570-fold.

By similar developments using Reaction 44 of Table 3-2 it can be shown that $Ag(HS)_2^-$ is not highly significant in the pe + pH range above 4.29 where α-Ag_2S precipitates. Its importance in soils can probably be discounted in view of the many other metal sulfide reactions that will occur in this low redox range (Lindsay, 1978).

3.3 CONCLUSIONS

The theoretical developments in this chapter show which silver minerals and complexes can form in chemical environments such as soils.

The oxides, sulfates, carbonates, and phosphates of Ag are too soluble to persist in soils. The minerals that could form, depending upon the activities of the halide ion, are silver iodide, silver bromide, and silver chloride. In addition, Ag(c) can form under reduced conditions and Ag_2S can eventually form if reduction approaches pe + pH of < 4.29. The exact conditions necessary for each of these minerals to precipitate are given.

Silver also forms many complexes in solution with halides, ammonia, nitrate, sulfate, sulfite, and bisulfide. The complexes with nitrate, sulfate, and silfite are not sufficiently stable to be significant in soils. The complexes with Br^-, Cl^-, and I^- are sufficiently stable to be significant. Their activity ranges are fully described. Complexes with ammonia are only slightly important even if rather large levels of NH_4^+ were to occur in soils. Under strongly reducing conditions (pe + pH < 4.7), $AgHS^o$ can become a significant complex for Ag^+.

The results of this development should be extremely useful in predicting the possible formation of silver minerals in soils and the solubility relationships associated with them, especially in relationship to the problem of silver-based nucleating agent impacts in a soil system.

Based on the available literature, silver ion is the major form of silver that has the ability to interact with biological systems. With the high levels of chlorides and other halides which might exist in typical soils, it can be expected that only a minor fraction of the total silver in soil solution would be present as free ionic silver.

Contribution from the Colorado State University Experiment Station published as Scientific Series Paper No. 2297. Supported in part by National Science Foundation RANN Weather Modification Program Project No. ENV73-07821 A03.

4

Inorganic Speciation of Silver in Natural Waters - Fresh to Marine

E. A. Jenne, D. C. Girvin, J. W. Ball, and J. M. Burchard

4.1 INTRODUCTION

An adequate evaluation of the impact of silver, from any point or nonpoint source such as cloud seeding, on surface-water ecosystems requires a knowledge of background levels of silver in surface and ground waters and of the quantity of anthropogenic silver being (or to be) added. Additionally, it is necessary to know the equilibrium distribution of silver among various aqueous complexes and solid phases, as well as the sorption potential of the various solute species for inorganic substrates and for biota. In order to predict effects on ecosystems, the accumulation of silver by various biota and the consequent toxicity effects need to be known for the pertinent solute species at the activity level which will occur. We know of no published study delineating the sorption potential of inorganic or biological substrates for the various individual solute silver species, whose charges vary from +1 through 0 to -3, although water "hardness" effects (e.g., hydroxide and carbonate complexation) have been observed (Davies, in press). Other elements are known to have vastly different biological potency and sorption potential for earth materials depending upon the dominant solute species (Jenne and Luoma, 1977), and silver should be no exception.

In this paper, analytical and thermodynamic data selected from the literature are used together with a comprehensive chemical model to investigate (1) the equilibrium distribution of silver among numerous inorganic solute complexes and solid phases, and (2) the manner in which this distribution changes during a transition from fresh-to-marine waters. The advantages and present limitations of such a modeling approach are discussed together with a statement of research needs for the clarification of geochemical and biological implications of the introduction of silver into water of varying chemical composition.

4.2 ACKNOWLEDGMENTS

We are pleased to acknowledge the assistance of John Conomos and David Peterson, of the U.S. Geological Survey in facilitating our shipboard E_H measurements and sulfide sampling as well as for providing the NO_3^-, NO_2^-, and NH_4^+ data used in the redox-potential calculations.

4.3 CHEMICAL MODEL

In natural waters, the distribution of solute metals among ligands is determined by a large number of interrelated and interdependent chemical reactions. Hence, in the calculation of the speciation of a single element among all or even a portion of the potentially important ligands, it is almost essential to utilize a comprehensive model capable of speciating many elements and ligands simultaneously. This study utilizes the prototype of a computerized chemical model known as WATEQ2 (Ball and Jenne, in prep.).

This model uses iterative solutions to mass action and mass balance equations until the mass balance conditions are obtained for input data. The earlier WATEQ (WATer EQuilibria) model of Truesdell and Jones (1973, 1974) has been expanded to include a large number of fluoride complexes (Nordstrom and Jenne, 1977), iron salts (D. K. Nordstrom, E. A. Jenne, and J. W. Ball, unpubl. data, 1976), iron and aluminum oxides and salts (E. A. Jenne, J. W. Ball, J. H. Barks, and J. M. Burchard, unpubl. data, 1977), and silver (this study).

The chemical model makes it possible to calculate readily the effect of changes in pH, temperature, salinity (e.g., ionic strength) and concentrations of specific cations and anions on the activity of uncomplexed silver and of the various aqueous silver complexes in the presence of other trace metals and major cations. This is of particular value when dealing with more than one toxic trace element for which synergistic biological effects occur.

The chemical model also utilizes the activity of various ions to compute activity products for numerous solid phases (e.g., $(A_{Ag^+})^2 \cdot (A_{S^{2-}})$ = activity product for Ag_2S). These activity products are divided by the solubility product constants, which are corrected to the temperature of the sample. The ratio of the activity product to the solubility product, along with appropriate uncertainty estimates for analytical and thermodynamic values, permits conclusions to be drawn as to whether the waters are supersaturated, in equilibrium, or undersaturated relative to various solid phases.

Some of the available log equilibrium constants and enthalpy-of-reaction data for Ag solutes and solids are compiled in Tables 4-1 and 4-2, respectively. The reactions indicated by an asterisk and the values underlined were incorporated into the chemical model. The free energy and enthalpy of formation for solute species and solid forms used in computing the log equilibrium constant and enthalpy of reaction data of Tables 4-1 and 4-2 are given in Table 4-3. Various solids which are too soluble to occur in nature are included in Table 4-2 for the sake of completeness but are not included in the model. Agreement of the thermodynamic data from different compilations and investigators is excellent, considering the diversity of the sources consulted. Association constants for zero ionic strength and 25°C are unavailable for several additional silver complexes. These complexes include $Ag_2SO_4^{o}$, $Ag_2I_3^{-}$, Ag_3I^{2+}, $NaAgI_3^{-}$, $Ag^{o}(aq)$, and $Me^{x+}(HS)_y$ where $Me^{x+} = Mn^{2+}$, Ca^{2+}, and Mg^{2+}, and y = 1, 2, 3, or 4. We assume that $AgHCO_3^{o}$ and $Ag_2CO_3^{o}$ complexes are unimportant. It is further assumed that thiosulfate* ($S_2O_3^{2-}$) and iodate (IO_3^{-}) concentrations are too low to be important in the estuarine water. Solubility data are not available for the sulfide minerals Ag_2S (amorphic), Ag_3AsS_3 (proustite), Ag_3SbS_3 (pyrargyrite), nor the mixed halogen Ag(Cl, Br, I). Solid-solution solubility data are not available, so far as we have been able to determine, for compounds of environmental interest except for Al-Si-OH (allophane) where the Al:Si ratio and solubility are pH dependent (Paces, 1973). The data are not yet at hand to permit an evaluation of the amount of silver complexed by dissolved organic substances of natural or anthropogenic origin. This lack of information places an important limitation upon the findings and conclusions from studies of the speciation of trace metals in natural waters.

The chemical composition for the marine and freshwater end members used in the model calculations are given in Table 4-4 and designated as Farallons and Rio Vista, respectively.† The composition at four intermediate salinities was obtained by simple mixing between end members. Temperature was incremented between the end members. A constant concentration of hydrogen sulfide (0.000 01 mg/L) and transition metals, except iron, were used (Table 4-4). Conservative mixing for all constituents is an implicit assumption. This assumption is, of course, not strictly valid. Insufficient analyses have been obtained to date to establish

*The $Ag(S_2O_3)_2^{2-}$ complex is reported to be the principal silver species released from photographic processing operations (Bard et al., 1976).

†Analyses for silver and selected transition metals are under way but concentration values were not available when the chemical modeling was done.

TABLE 4-1. *Compilation of log equilibrium constant and enthalpy of reaction data for silver solute forms (at 25°C and zero ionic strength, calculations for given reaction are as follows:* $\Delta G_r^o = \Delta G_f^o$ *(Σ products) —* ΔG_f^o *(Σ reactants),* $\log_{10} K_r^o = \Delta G_r^o / -2.303\ RT$*, where* $-2.303\ RT = -1.364$ *at 25°C, and* $\Delta H_r^o = \Delta H_f^o$ *(Σ products) –* ΔH_f^o *(Σ reactants). [Asterisk leading the reaction signifies the mineral phase was entered in the model, underlined values are those selected for use in the model.]*

Reactions	Wagman et al. (1969) log K_r^o	Wagman et al. (1969) ΔH_r^o	Naumov et al. (1974) log K_r^o	Naumov et al. (1974) ΔH_r^o	Helgeson (1969) log K_r^o	Helgeson (1969) ΔH_r^o	Other Sources log K_r^o	Other Sources ΔH_r^o	Reference
*$Ag^+ + Br^- \rightleftharpoons AgBr^o$	$\underline{4.24}$[2]								
*$Ag^+ + 2Br^- \rightleftharpoons AgBr_2^-$	$\underline{7.28}$								
*$Ag^+ + 3Br^- \rightleftharpoons AgBr_3^{2-}$	$\underline{8.71}$								
$Ag^+ + Br^- + 3Cl^- \rightleftharpoons AgCl_3Br^{3-}$	7.89								
$Ag^+ + 3Br^- + Cl^- \rightleftharpoons AgClBr_3^{3-}$	9.46								
*$Ag^+ + Cl^- \rightleftharpoons AgCl^o$	$\underline{3.27}$	$\underline{-2.68}$	3.31	-2.70	3.31		2.69		Yatsimirskii & Vasil'ev (1960)
							3.27		Seward (1976)
*$Ag^+ + 2Cl^- \rightleftharpoons AgCl_2^-$	5.27	$\underline{-3.93}$	5.25	-3.90	5.24		4.75		Yatsimirskii & Vasil'ev (1960)
							$\underline{5.18}$		Seward (1976)
*$Ag^+ + 3Cl^- \rightleftharpoons AgCl_3^{2-}$					$\underline{5.29}$				
*$Ag^+ + 4Cl^- \rightleftharpoons AgCl_4^{3-}$					$\underline{5.51}$		5.92		Yatsimirskii & Vasil'ev (1960)
*$Ag^+ + F^- \rightleftharpoons AgF^o$	$\underline{0.36}$	$\underline{-2.83}$	0.38	-2.84					
*$Ag^+ + HS^- \rightleftharpoons AgHS^o$			$\underline{14.05}$						

TABLE 4-1 *(continued)*

$*Ag^+ + 2HS^- \rightleftharpoons Ag(HS)_2^-$			18.45				
$*Ag^+ + I^- \rightleftharpoons AgI^o$	6.6						
$*Ag^+ + 2I^- \rightleftharpoons AgI_2^-$	10.68						
$*Ag^+ + 3I^- \rightleftharpoons AgI_3^{2-}$	13.37						
$*Ag^+ + 4I^- \rightleftharpoons AgI_4^{3-}$	14.08						
$*Ag^+ + H_2O \rightleftharpoons AgOH^o + H^+$	-11.91		-11.99			-12.0	Baes & Mesmer (1976)
$*Ag^+ + 2H_2O \rightleftharpoons Ag(OH)_2^- + 2H^+$	-24.00		-24.00			-24.0	Baes & Mesmer (1976)
$*Ag^+ + SO_4^{2-} \rightleftharpoons AgSO_4^-$	1.29	1.49	1.30	1.49	1.31		
$*Ag^+ + NO_3^- \rightleftharpoons AgNO_3^o$	-0.29						
$*Ag^+ + 2NO_2^- \rightleftharpoons Ag(NO_2)_2^-$	2.22						
$Ag^+ + S_2O_3^{2-} \rightleftharpoons AgS_2O_3^-$			8.60				
$Ag^+ + 2S_2O_3^{2-} \rightleftharpoons Ag(S_2O_3)_2^{3-}$			13.43	-19.05			
$Ag^+ + SO_3^{2-} \rightleftharpoons AgSO_3^-$	5.37						
$Ag^- + 2SO_3^{2-} \rightleftharpoons Ag(SO_3)_2^{3-}$		-28.73				8.477	Gedansky & Hepler (1969)
$Ag^+ + NH_3^o \rightleftharpoons AgNH_3^+$	3.315						
$Ag^+ + 2NH_3^o \rightleftharpoons Ag(NH_3)_2^+$	7.222	-13.454					
$Ag^+ + 2NH_3^o + Cl \rightleftharpoons Ag(NH_3)_2Cl^o$	0.58						
$Ag^+ + 2NH_3^o + Br^- \rightleftharpoons Ag(NH_3)_2Br^o$	4.356						

TABLE 4-2. *Compilation of log equilibrium constant and enthalpy of reaction for silver solid forms (at 25°C and zero ionic strength). [Asterisk leading the reaction signifies the mineral phase was entered in the model, underlined values are those selected for use in the model.]*

Mineral	Reaction	Wagman et al. (1969)		Naumov et al. (1971)		Garrels & Christ (1965)		Other Sources		Reference
		$\log K_r^\circ$	ΔH_r°	$\log K_r^\circ$	ΔH_r°	$\log K_r^\circ$	ΔH_r°	$\log K_r^\circ$	ΔH_r°	
	$^*Ag \rightleftharpoons Ag^+ + e^-$	$\underline{-13.511}$	$\underline{25.234}$							
bromyrite	$^*AgBr \rightleftharpoons Ag^+ + Br^-$	$\underline{-12.27}$	$\underline{20.17}$	-12.302	20.284	-12.304	20.190	-12.301		Gedansky & Hepler (1969)
								-12.301		Robie & Waldbaum (1968)
cerargyrite	$^*AgCl \rightleftharpoons Ag^+ + Cl^-$	$\underline{-9.752}$	$\underline{15.652}$	-9.749	15.654	-9.752	15.649	-9.73		Seward (1976)
								-9.73		Robie & Waldbaum (1968)
	$^*Ag_2CO_3 \rightleftharpoons 2Ag^+ + CO_3^{2-}$	$\underline{-11.07}$	$\underline{9.53}$	-11.080	9.518	-11.083	9.960	-11.097		Gedansky & Hepler (1969)
	$AgF{\cdot}2H_2O \rightleftharpoons Ag^+ + F^- + 2H_2O$	0.87	0.50							
	$^*AgF{\cdot}4H_2O \rightleftharpoons Ag^+ + F^- + 4H_2O$	$\underline{0.55}$	$\underline{4.27}$							
iodyrite	$^*AgI \rightleftharpoons Ag^+ + I^-$	$\underline{-16.07}$	$\underline{26.82}$	-16.066	26.224	-16.074	26.850			
	$AgNO_3 \rightleftharpoons Ag^+ + NO_3^-$	0.13	5.40	0.157	5.400	0.227	5.368			
	$AgNO_2 \rightleftharpoons Ag^+ + NO_2^-$	-3.65	11.0							
	$^*\beta Ag_2O + 2H^+ \rightleftharpoons 2Ag^+ + H_2O$	$\underline{12.56}$	$\underline{-10.43}$	12.574	-10.43			12.596		Gedansky & Hepler (1969)
								12.58		Baes & Mesmer (1976)
	$Ag_2O_2 + 4H^+ \rightleftharpoons 2Ag^+ + 2H_2O + 2e^-$	60.92	-78.36							
acanthite[1]	$^*\beta Ag_2S + H^+ \rightleftharpoons 2Ag^+ + HS^-$	$\underline{-36.05}$	$\underline{53.3}$	-36.053	53.298	-36.085	53.410	-37.9		Gedansky & Hepler (1969)
								-37.9		Robie & Waldbaum (1968)
	$^*Ag_2SO_4 \rightleftharpoons 2Ag^+ + SO_4^{2-}$	$\underline{-4.92}$	$\underline{4.25}$	-4.945	4.148	-4.904	4.220			
	$^*Ag_3PO_4 \rightleftharpoons 3Ag^+ + PO_4^{3-}$	-15.98								
	$AgOH + H^+ \rightleftharpoons Ag^+ + H_2O$					11.93				

[1]Low temperature form (below 177°C) which is referred to as the β form by some investigators and as α form by others. Naumov et al. (1974), indicate the data of theirs which we have used here are for argentite (the high temperature form). However, we suspect that they have transposed the mineral forms in their report.

the magnitude of the error introduced by this assumption. The small differences in molalities of some constituents between the end members result from density differences of the waters.

The concentration of silver was also held constant (0.000 040 mg/liter). After a critical evaluation of background levels of silver in natural waters (D. C. Girvin and E. A. Jenne, unpubl. data) this value was chosen as representative of coastal, estuarine, and Sacramento River waters.

There are small but finite compositional differences between Farallons water and seawater (Table 4-4). For the purposes of these model calculations, however, we will refer to the Farallons as the marine end member. The utility and meaning of salinity, i.e., its constant proportionality to chloride ion and other major cation and anion concentrations does not, of course, apply at low salinities. Salinity is used in the text and included in the plots as a convenient frame of reference.

TABLE 4-3. *Free energy and enthalpy of formation for solute species and solid forms used in computing the log equilibrium constant and enthalpy of reaction data of tables 1 and 2.*[1]

Solute Species or Solid Form	ΔG_f^o	ΔH_f^o	Reference
	A. Silver Solute Species		
Ag^+	18.433[2]	25.234	Wagman et al. (1969)
$AgBr^o$	-12.2	—	Wagman et al. (1969)
$AgBr_2^-$	-41.2	—	Wagman et al. (1969)
$AgBr_3^{2-}$	-68.0	—	Wagman et al. (1969)
$AgCl^o$	-17.4	-17.4	Wagman et al. (1969)
$AgCl_2^-$	-51.5	-58.6	Wagman et al. (1969)
$AgCl_3^{2-}$	-82.85	-97.82	Helgeson (1969)
$AgCl_4^{3-}$	-114.58	-137.17	Helgeson (1969)
AgF^o	-48.7	-57.1	Wagman et al. (1969)
$AgHS^o$	2.1	—	Naumov et al. (1974)
$Ag(HS)_2^-$	-1.0	—	Naumov et al. (1974)
AgI^o	-2.9	—	Wagman et al. (1969)
AgI_2^-	-20.8	—	Wagman et al. (1969)
AgI_3^{2-}	-36.8	-43.5	Wagman et al. (1969)
AgI_4^{3-}	-50.1	—	Wagman et al. (1969)
$AgOH^o$	-22.0	—	Wagman et al. (1969)

(continued)

[1]These values are taken preferentially from Wagman et al. (1968, 1969). The other values were taken from Helgeson (1969) or Naumov et al. (1974), as necessary.

[2]Helgeson (1969) takes his ΔG_f^o from Latimer (1952) and his ΔH_f^o from Wagman et al. (1969). Naumov et al. (1974) take both values from Wagman et al. (1969).

TABLE 4-3. (continued)

Solute Species or Solid Form	ΔG_f^o	ΔH_f^o	Reference
$Ag(OH)_2^-$	-62.2	—	Wagman et al. (1969)
$AgSO_4^-$	-161.3	-190.6	Wagman et al. (1969)
$AgNO_3^0$	-7.78	—	Wagman et al. (1969)
$Ag(NO_2)_2^-$	-2.4	—	Wagman et al. (1969)
	B. Silver Solid Forms		
AgBr	-23.16	-23.99	Wagman et al. (1969)
AgCl	-26.244	-30.370	Wagman et al. (1969)
Ag_2CO_3	-104.4	-120.9	Wagman et al. (1969)
$AgF \cdot 4H_2O$	-274.2	-331.8	Wagman et al. (1969)
AgI	-15.82	-14.78	Wagman et al. (1969)
βAg_2O	-2.68	-7.42	Wagman et al. (1969)
βAg_2S	-9.43	-7.03	Wagman et al. (1969)
Ag_2SO_4	-147.82	-171.10	Wagman et al. (1969)
Ag_3PO_4	-210		Wagman et al. (1969)
	C. Other Solute Species		
Br^-	-24.84	-29.05	Wagman et al. (1968)
Cl^-	-31.372	-39.952	Latimer (1952), Wagman, et al. (1968)
CO_3^{2-}	-126.17	-161.84	Wagman et al. (1968)
F^-	-66.64	-79.50	Wagman et al. (1968)
H_2O	-56.687	-68.315	Wagman et al. (1968)
HS^-	2.88	-4.2	Naumov et al. (1974)
I^-	-12.33	-13.19	Wagman et al. (1968)
NO_3^-	-26.60	-49.56	Wagman et al. (1968)
NO_2^-	-8.9	-25.0	Wagman et al. (1968)
PO_4^{3-}	-243.5	-305.3	Wagman et al. (1968)
SO_4^{2-}	-177.97	-217.32	Wagman et al. (1968)

4.4 SPECIATION

The variation of activity of the soluble silver complexes with salinity from fresh (162 μmho at 25°C) to marine waters is given in Figure 1. As the salinity decreases from 33.5 to approximately 0.13 o/oo (total grams dissolved salt per kilogram of water) the activity of uncomplexed silver (A_{Ag^+}) increases sharply (from 3×10^{-15} to 4×10^{-11})* while the activities of all

*Activity has the dimension of molality; it is defined as the molal concentration times dimensionless activity coefficient (Lewis and Randall, 1961, p. 251).

chloride complexes, except $AgCl^{o}$, decrease. For marine waters $A_{AgCl_2}{}^{-} > A_{AgCl_3}{}^{2-} > A_{AgCl_4}{}^{3-} > A_{AgCl^{o}}$.

One of the significant uses of a comprehensive chemical model which is applicable to natural waters is to evaluate the need for additional analytical parameters. Thus, the need for bromide, iodide and sulfide analytical data was evaluated in the absence of any known analyses for these constituents in these waters. The bromide and iodide concentrations used and the sources of these estimates are shown in Table 4-4. At the concentrations of bromide, iodide, and chloride in marine and representative fresh waters, bromide and iodide complexes account for only a minor portion of the dissolved silver (Figure 4-1B).

The model was next used to perform a sensitivity analysis as to the relative importance of the chloride, bromide and iodide complexes, for the case where the concentration estimates for bromide and iodide used were low by as much as a factor of 10. This possibility was tested by alternately increasing bromide and iodide 10-fold. The results (Figure 4-1C) indicate that analytical measurements for these two constituents would not be worthwhile. Even when the input bromide concentration is increased 10-fold (Figure 4-1C) the activity of the most abundant bromide complex ($AgBr_2^-$) in the marine end member is only 1.7% of the activity of most abundant chloride complex ($AgCl_2^-$) and is equivalent to the activity of the least abundant chloride complex ($AgCl^{o}$). The iodide complexes are still quantitatively unimportant even when the input iodide concentration is increased 10-fold. The probable occurrence of 25-33% of the iodine as iodate ion (Riley and Chester, 1971, p. 78) would further reduce the importance of silver iodide complexes.

A potentially important inorganic ligand missing from the foregoing discussion is sulfide. Hydrogen sulfide is such an unstable constituent that its probable concentration is much more difficult to assess than that of bromide or iodide. We first considered what the dominant redox couple(s) might be. Platinum electrode redox potentials measured in July 1976 (Table 4-6) ranged from 0.36 V at the low salinity end to 0.39 V at the Golden Gate Bridge near the high salinity end of the estuary transect. In accordance with the observations of Mortimer (Golterman and Clymo, 1970, p. 148) listed in Table 4-5, these platinum electrode redox potentials (Table 4-6) suggest that NO_2^-/NH_4^+ would be the qualitatively dominant couple. In contrast, the observed dissolved oxygen values (Table 4-4) of near saturation would indicate NO_3^-/NO_2^- as the qualitatively important couple (Table 4-5). These couples are termed "qualitatively" important because it is the product of the concentrations of a couple and its electron exchange rate that determines its quantitative importance.

The above observations indicate that the various redox couples are not at equilibrium in these waters. This may be the general case in open bodies of water. In contrast to the findings of Mortimer (Golterman and

TABLE 4-4. *Chemical composition of seawater and endmember waters used in chemical modeling. [Molalities given in "E-format," e.g., for E^{-2} read 10^{-2}.]*

Component	Seawater[1] mg/liter	Seawater[1] molality	Farallons[2] mg/liter	Farallons[2] molality	Rio Vista[3] mg/liter	Rio Vista[3] molality	Source
Ca	412	1.06E-02	394	1.02E-02	12.2[a]	3.04E-04	[1]Brewer, p. 417-419 *in* Riley &
Mg	1290	5.50E-02	1235	5.26E-02	7.5[a]	3.08E-04	Skirrow (1975), unless otherwise
Na	10770	4.85E-01	10313	4.64E-01	12.0[a]	5.22E-04	indicated.
K	380	1.01E-02	382	1.01E-02	1.4[a]	3.58E-05	[2]Calculated from salinity of sample
Cl	18800	5.49E-01	18530	5.41E-01	9.9[f]	2.79E-04	taken March, 1976, unless otherwise
SO_4	2709	2.92E-02	2579	2.80E-02	7.7[f]	8.02E-05	indicated. Actual analyses for specific
HCO_3	140	2.38E-03	140[d]	2.37E-03	75.2[d]	1.23E-03	components are also for sample taken
Fe^{2+}	0.0029[ab]	7.68E-08	0.0029[a]	5.37E-08	0.015[a]	2.78E-07	March, 1976.
Fe^{3+}	.0029[a]	5.38E-08	0.0029[a]	5.37E-08	0.0007[a]	1.25E-08	[3]Livingstone, 1963, unless otherwise
H_2S	0.00001[c]	3.04E-10	0.00007[f]	3.04E-10	.00001[c]	2.94E-10	indicated.
CO_3	0	0	0	0	0	0	[a]Unpub. data for Al, Ag: Ball et al., 1972;
SiO_2	4.26	7.34E-05	0.520[d]	8.96E-06	17.04[d]	2.84E-04	Fe: Vivit, value for Fl (Farallons)
NH_4	0	0	0.0029[d]	1.66E-07	0.144[d]	7.98E-06	used for seawater to provide an
B	4.44	4.25E-04	5.15[a]	4.93E-04	0.050[a]	4.63E-06	E_H estimate;
PO_4	0.190	2.07E-06	0.089[d]	9.70E-07	0.210[d]	2.21E-06	B: Ball;
Al	0.002	7.68E-08	0.00213[e]	8.17E-08	0.005[a]	1.85E-07	DOC: Girvin;

TABLE 4-4 *(continued)*

F	1.30	7.09E-05	1.24	6.75E-05	0.10[f]	5.26E-06	Ca, Mg, Na, K: Vivit
NO_3	0	0	0.617[d]	1.03E-05	0.898[d]	1.45E-05	[b]Riley & Skirrow (1975) give value of
NO_2	0	0	0.016[d]	3.62E-07	0.019[d]	4.20E-07	0.002 mg/liter for Fe total.
Br	67.0	8.68E-04	64.4	8.34E-04	0.006	7.51E-08	[c]Analytical value, min. established,
I	0.06	4.90E-07	0.06	4.89E-07	0.0018	1.42E-08	Ball.
Ag	0.0004[g]	3.84E-10	0.0004[g]	3.84E-10	0.00004[a]	3.71E-10	[d]Peterson, D., person. communication.
Mn	0.0044[h]	8.30E-08	0.0044[h]	8.29E-08	0.0044[a]	8.01E-08	[e]Incremented on salinity gradient.
Cd	0.0001	9.22E-10	0.0001[l]	9.21E-10	0.0001[l]	8.90E-10	[f]Analysis, U.S. Geol. Survey, Central
Cu	0.0005	8.15E-09	0.0005[l]	8.14E-09	0.0005[l]	7.87E-09	Lab, Denver, Colo.
Ni	0.0018[h]	3.18E-08	0.0018[h]	3.17E-08	0.0018[l]	3.07E-08	[g]Estimate for near shore marine waters
Pb	0.00003	1.50E-10	0.00003[l]	1.50E-10	0.00003[l]	1.45E-10	(Girvin & Jenne, in prep.).
Zn	0.0049	7.76E-08	0.0049[l]	7.76E-08	0.0049[l]	7.50E-08	[h]Carvajal & Landergren (1969).

F = Farallons

RV = Rio Vista

	F	RV
Diss. Oxygen	9.50[d]	10.94[d]
Diss. Organic Carbon	2.6[a]	2.5[a]
Salinity °/00	33.480[d]	0.130[d]
Conductivity (μmhos)		162[f]
Temperature (°C)	10.6[d]	9.5[d]
pH	8.19[d]	8.01[d]

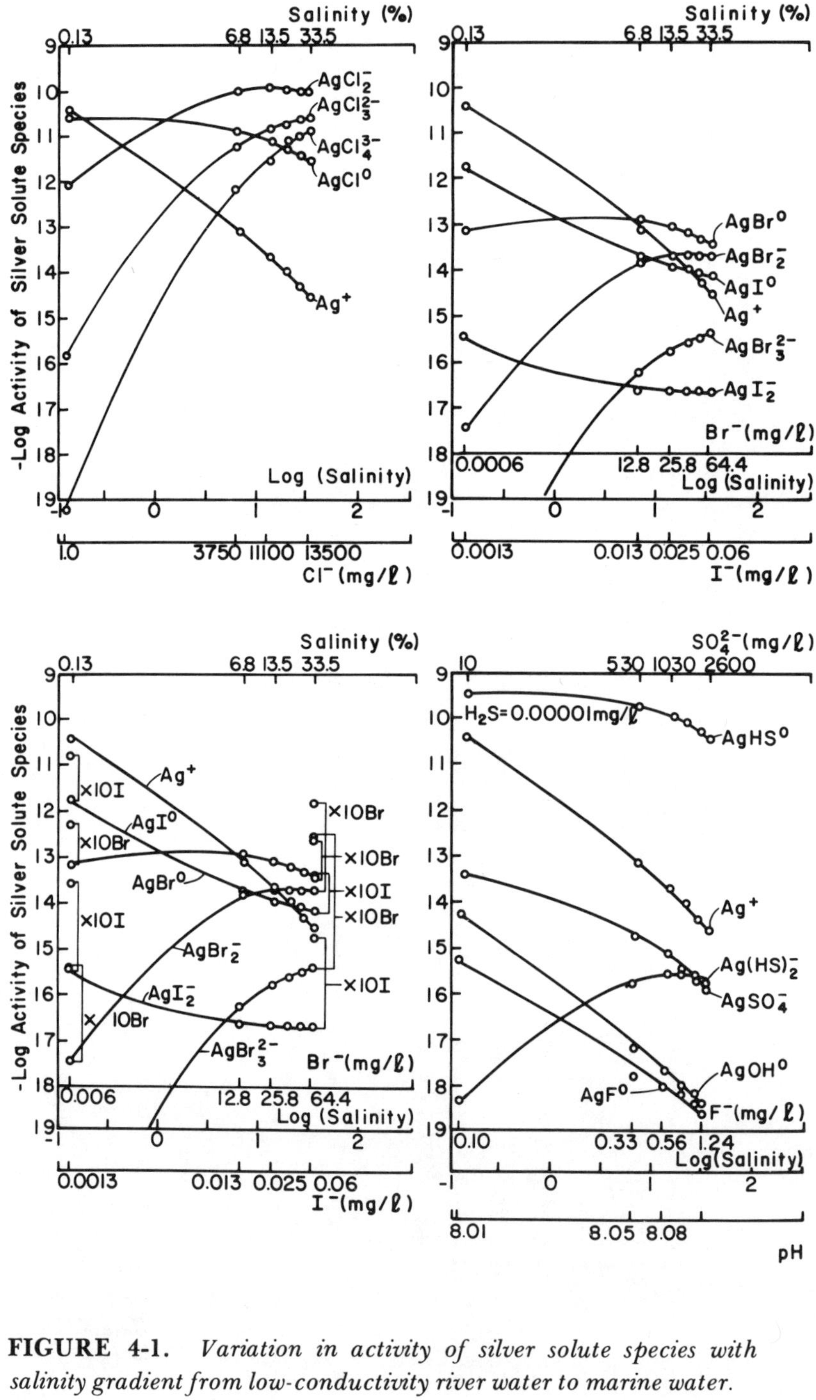

FIGURE 4-1. *Variation in activity of silver solute species with salinity gradient from low-conductivity river water to marine water.*

Clymo, 1970) given in Table 4-5, others have suggested that NO_3^- is reduced to NO_2^- to a significant extent only below dissolved oxygen levels of 1 mg/liter (Inter. Rice Res. Instit., 1964, p. 203). The question then arises as to whether geochemically significant levels of dissolved hydrogen sulfide could occur in San Francisco Bay waters.

Yoshimura and Wada (1938) noted that in a brackish lake dissolved hydrogen sulfide became detectable (~0.2 mg/liter) only at dissolved oxygen levels below 1 mg/liter, but also that 1.5 mg/liter of hydrogen sulfide coexisted with 0.50 mg/liter of dissolved oxygen. Bella (1975) found dissolved sulfide in the open waters over the tidal flats of a shallow Oregon estuary to range from 0.1 to >10 mg/liter. He found as much as 1 mg/liter of sulfide coexisting with 4 to 6 mg/liter of dissolved oxygen. The differences between these two studies may be that in the shallow tidal flats the rates of sulfide production and tidal dispersion into overlying waters are much greater than the sulfide oxidation rate. In the deeper water body studied by Yoshimura and Wada, diffusive and advective mixing rate may be much smaller than sulfide oxidation rate.

San Francisco Bay is typically well mixed, as indicated by the absence or near absence of vertical salinity, dissolved oxygen, and dissolved silica gradients (David Peterson, pers. comm., 1977). In general, dissolved oxygen in surface and bottom (10 m) waters are 95 to 85% of saturation, respectively. Short-lived periods of partial stratification do occur, however, during periods of peak river discharge. Bottom sediment is probably a major source of hydrogen sulfide generation and release to the water column, especially over the extensive areas of fine-grained tidal flats. Near the completion of this study, a report by Silvey and Irwin (1969) was found which states that dissolved hydrogen sulfide levels of 1 to 6 mg/liter had been found in the overlying waters of one mud flat area in northern San Francisco Bay. Their report does not indicate dissolved oxygen levels and gives no tabular data for dissolved hydrogen sulfide.

The reaction of dissolved hydrogen sulfide and oxygen is complex in that polysulfides, thiosulfates, and sulfites are probably formed and the

TABLE 4-5. *Predominant redox couples, prevalent redox potentials, and prevalent dissolved-oxygen concentrations (after Golterman and Clymo, 1969, and Mortimer, 1942).*

Redox Couple	Redox Potentials[1]	Dissolved Oxygen
	. . . . volts	mg/liter
$NO_3^- \rightleftharpoons NO_2^-$	0.45 to 0.40	4
$NO_2^- \rightleftharpoons NH_4^+$	0.40 to 0.35	0.4
$Fe^{3+} \rightleftharpoons Fe^{2+}$	0.30 to 0.20	0.1
$SO_4^{2-} \rightleftharpoons S^{2-}$	0.10 to 0.06	0.0

[1]The lower value is the limit below which the oxidized species could not be detected.

TABLE 4-6. *The range of redox potential values calculated from activities of participating species for the salinity gradient used in this study and from platinum electrode measurements at the Bay Bridge and Rio Vista. [The March 1976 dissolved oxygen values ranged from 8.6 to 10.9 mg/liter.]*

Redox Couple	Redox Potentials
	. . . . volts. . . . .
$NO_3^- \rightleftharpoons NO_2^-$	0.44 to 0.43
$NO_3^- \rightleftharpoons NH_4^+$	0.33 to 0.32
$Fe^{3+} \rightleftharpoons Fe^{2+}$	0.12 to 0.05
$SO_4^{2-} \rightleftharpoons H_2S^o$	-0.17 to -0.20
Pt electrode	0.39 to 0.36

oxidation of hydrogen sulfide may be catalyzed by metallic ions and accelerated by some organic substances (Bella, 1975) and suspended particulates (Wheatland, 1954). Bella (1975) also noted that independent investigators described the oxidation as a second order reaction with respect to dissolved oxygen and hydrogen sulfide, even though temperature, pH, dissolved organic substrates, and initial concentrations of oxygen and hydrogen sulfide all affect the rate of oxidation. The hydrogen sulfide half-life in oxygenated seawater is reported to vary between minutes and days. Thus, the available literature data would suggest that hydrogen sulfide levels of fractional milligrams per liter could occur in the oxygenated San Francisco Bay estuary waters, but that variable production and mixing rates as well as the paucity of data preclude the use of literature values.

Analytical work is underway to obtain determinations of total dissolved hydrogen sulfide values in San Francisco Bay waters. An initial estimate of total aqueous dissolved hydrogen sulfide for the vicinity of the Golden Gate Bridge is 0.000 01 mg/liter using a provisional analytical method (J. W. Ball and E. A. Jenne, unpubl. data, 1977). This value was used in the entire set of mathematical mixtures.

The silver bisulfide ion pair (Figure 4-1D), at the hydrogen sulfide level used (0.000 01 mg/liter), is the dominant silver species in the entire transect. The fluoride, hydroxide, sulfate, and double bisulfide of silver are all trivial quantities (Figure 4-1D) under these conditions as is $A_{Ag}{}^{2+}$ (not shown). The activity of the silver-ammonia complex was calculated and also found to be negligible.

An arbitrary 100-fold reduction in the input hydrogen sulfide concentration reduces the activity of $AgHS^o$ by 50- and 100-fold for the marine and river end members, respectively (Figure 4-2). At this lower hydrogen sulfide concentration, $AgHS^o$ is not the dominant solute silver complex (compare Figure 4-1A and D with Figure 4-2). Ionic silver is the predominant freshwater form of dissolved silver while the chloride complexes pre-

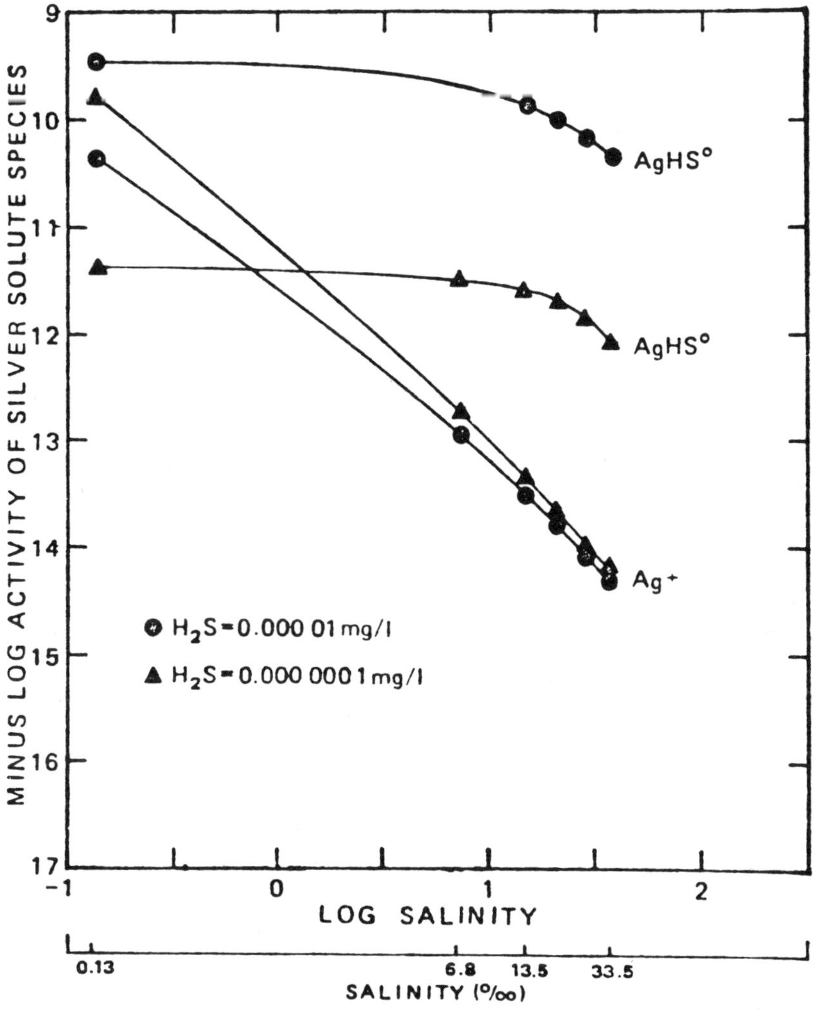

FIGURE 4-2. *Variation in activity of Ag^+ and $AgHS^o$ at two sulfide levels through the salinity gradient.*

dominate in marine waters. It is quite obvious that accurate analytical measurements of hydrogen sulfide levels are required to reliably compute aqueous speciation of silver. The sulfide complexes of silver and other trace metals are likely to be even more important in lakes and impoundments where sulfide levels may reach tens and even hundreds of milligrams per liter (Gaines et al., 1972).

A significant uncertainty exists in the estimation of activity coefficients for uncharged solute species (e.g., $AgCl^o$, etc.). All activity coefficients for uncharged solutes have been calculated in this study from the empirical relationship $\Upsilon_{mel}{}^{o} = 10^{0.1I}$, where I is the ionic strength (Truesdell and Jones, 1973). The use of the relation $\gamma_{mel}{}^{o} = 10^{0.06I}$ (Reardon and Langmuir, 1974) has negligible effect on the neutral solute species activity coefficients at Rio Vista but reduces the coefficient by about 8% at the Farallons. The activity coefficient for neutral species calculated with the $10^{0.1I}$ factor range from 1.00 at Rio Vista to 1.16 at the Farallons. Reardon and Langmuir (1974) point out that for certain neutral solutes (e.g., $MgCO_3{}^{o}$), which are likely to be separated by a sheath of water molecules, the activity coefficient is likely to be reduced to about that of a monovalent ion. We, therefore, recomputed the speciation for the end member waters setting the activity of neutral species equal to that of Na^+ ($\gamma_{Na}{}^{+} = 0.949$ at Rio Vista and 0.749 at the Farallons). The net effect (see Appendix) is to decrease the activity of most silver solute species by 8 to 10% but this does not affect the relative quantity of the solute species shown in Figure 4-1. The effect on the AP/K ratios is generally 0.1 or less.

The quantity of silver in association with low concentrations of strong ligands may vary in the presence of other metals. To study the influence of other metals on the chloride, sulfide, iodide, and bromide complexes of silver, we recalculated the speciation of silver in the absence of the following metals: Fe^{3+}, Fe^{2+}, Mn^{2+}, Cd^{2+}, Cu^{2+}, Ni^{2+}, Pb^{2+}, and Zn^{2+} (concentrations given in Table 4-4). At the total dissolved sulfide level used (3×10^{-10} molal), the omission of these trace metals (total quantity $= 3.38 \times 10^{-7}$ molal) has little or no effect on the activity of various silver solute species. The omission of these trace metals increases the activity of the various complexes shown in Figure 4-1 by a maximum of one-half the width of the points plotted. It should be noted that the seawater concentrations of Cd, Cu, Ni, and Zn cited in Table 4-4 (Brewer, 1975), and used in these calculations, are significantly higher than those currently being observed by a number of investigators utilizing great care to avoid contamination.

4.5 MINERAL EQUILIBRIA

The low activity of silver in these waters would seem to preclude any solubility control on silver. Krauskopf (1956) noted that silver levels in marine waters were below saturation with probable solid forms such as AgCl. None of the solid forms of silver which are included in the model appear to control dissolved silver levels. However, at the hydrogen sulfide concentration used (0.000 01 mg/liter) the apparent oversaturation with $Ag_2S(s)$ ranges from 7 to 12 orders of magnitude. This seems unlikely to be real and is outside the range of errors attributable to the thermodynamic and/or analytical data. Although there could be kinetic inhibitions to silver sulfide precipitation, the widespread use of H_2S precipitation in qualitative and quantitative chemistry make this possibility unattractive. We suspect the problem is simply one of utilizing thermodynamic data for an inappropriate solid phase. The thermodynamic data used were obtained by calorimetry and from high-temperature equilibrium and emf measurements on a presumably high crystalline silver sulfide phase* instead of an amorphic variety which would be expected to precipitate from surface waters. Not only is the estuarine precipitate likely to be amorphic but the stoichiometry may be different from that of the solid for which thermodynamic data were obtained. Similar computations indicate that these waters are oversaturated with respect to the crystalline iron sulfide (pyrite) by 3 to 12 orders of magnitude. However, they are undersaturated with amorphic iron sulfide by 6 to 8 orders of magnitude. Additionally, these waters are near equilibrium with amorphic ferric hydroxide but are oversaturated by about 6 and 13 orders of magnitude for the crystalline iron oxides, lepidocrocite and hematite, respectively. Both the crystallinity and stoichiometry are different for the two iron sulfides and the two oxides. The combination of these two factors plus particle size effects are likely to be the cause of the large apparent oversaturation with $Ag_2S(s)$. Clearly, thermodynamic data on amorphic $Ag_2S(s)$ are required before deciding if it is an equilibrium solubility phase in estuaries.

The apparent oversaturation with respect to $Ag_2S(s)$ in conjunction with the high activity of $AgHS^o$ led us to check the adequacy of the model for HS^- complexes of "major" cations. In checking Sillen and Martell (1964) and other sources, no thermodynamic data were found for Mn^{2+}, or Al^{3+} complexes of HS^-. Association constants for $Fe^{2+}(HS)_2^o$ and $Fe^{2+}(HS)_3^-$ were added to the model but their effect was trivial in that their inclusion had no measurable effect on the A_{AgHS^o} (5.180×10^{-11}) for water near Farallons.

It is likely that a significant portion of the solute silver is present as organic complexes and compounds. However, it does not appear reasonable that organic complexing accounts for more than a portion of the apparent oversaturation.

At Rio Vista, the water apparently is oversaturated with respect to Ag(s). Superficially, the oversaturation with Ag(s) at Rio Vista could be ascribed to the high A_{Ag^+}, due to the low chloride levels in the river water compared to the estuary. However, this seems unreasonable in that the oversaturation would presumably occur up river as well. On the other hand, use of redox potential values from platinum electrode measurements (0.36 V) suggest that the river water may well be far undersaturated with respect to Ag(s). Likewise, use of redox potentials calculated from analytical measurements for the couples NO_3^-/NO_2^- (0.44 V) and

*The data used is for monoclinic Ag_2S (acanthite) which is the stable form up to 177°C (Robie and Waldbaum, 1968).

NO_3^-/NH_4^+ (0.32 V) indicates undersaturation with Ag(s). These calculations emphasize redox potential considerations are very important to meaningful calculation of Ag(s) equilibrium.

Gang and Langmuir (1974) found Ag in several of the ground waters they examined to be near equilibrium with Ag(s), but noted that uncertainties in the analytical determination of silver concentrations and platinum electrode redox potentials prevented more definitive conclusions from being made. They did not determine H_2S levels in their ground waters.

4.6 DISCUSSION

Aqueous speciation is necessary in order to make meaningful tests of solubility equilibrium hypotheses. Aqueous speciation may also be necessary to interpret the adsorption onto inorganic substrates as well as biological uptake. Difficulties arise if chemical modeling efforts consider an inadequate number of trace elements and complexes simultaneously as do most of the current models.

For example, the quantitative importance of the $AgHS^o$(aq) complex relative to that of the monochloride complex as well as the increased portion of silver which is contained in the bichloride and trichloride silver complexes at near maximum estuarine salinities illustrates the need to consider all of the complexes which could constitute a significant portion of the dissolved silver. Obviously, computations excluding these or other appropriate complexes will lead to erroneous predictions of the dominant complexes and to erroneous conclusions concerning the nonequilibrium nature of natural waters.

For the Farallons end member, an input hydrogen sulfide concentration of 0.000 000 1 mg/liter (3×10^{-12} molal) results in a computed $A_{H_2S^o}$ of 1.70×10^{-13}, A_{HS^-} of 2.87×10^{-12} and $A_{S^{2-}}$ of 4.55×10^{-17}. The lower $A_{H_2S^o}$, as compared to the input concentration of hydrogen sulfide, is due to the dissociation reaction $H_2S \rightleftharpoons HS^- + H^+$ and to the complexation by HS^- of the various metals.

In order to adequately compute speciation, activity coefficients must also be considered. Activity coefficients for charged species are generally computed by the extended Debye-Hückel or the Davies equation (Davies, 1962). The latter was used in this study because it may be more accurate at higher dissolved solid levels and does not require, as does the Debye-Hückel equation, ion size parameters, which are unknown for most trace element complexes. In the Farallons samples, the activity coefficients are 1.16, 0.75, 0.32, and 0.074 for $AgCl^o$, $AgCl_2^-$, $AgCl_3^{2-}$, and $AgCl_4^{3-}$, respectively. Thus, the concentrations of the silver chloride species are appreciably different from their activities.

A comprehensive chemical model capable of the speciation of inorganic silver in waters ranging from very low dissolved solids to sea water has several environmentally important uses. One example is to compute the activities of potentially toxic complexes of trace elements such as silver. Then, toxicity can be interpreted in terms of the activity of particular solute species, presumably the uncomplexed ion (Jenne and Luoma, 1977), as functions of salinity and trace element concentrations.

In the south San Francisco Bay the concentration of dissolved silver is approximately 0.000 080 mg/liter (D. C. Girvin and M. E. Tatro, unpubl. data, 1977). At least part of this silver apparently exists in a form that is highly available biologically, as is indicated by the relatively high silver concentrations of 80 and 196 mg/Kg (dry weight) in the eastern soft shell clam (*Mya arenaria*) and the Pacific oyster (*Crassistrea gigas*) (Girvin et al., 1975, p. 58 and 63). These South Bay organisms were collected from waters with salinities of between 22 and 24 o/oo, but the salinity in the south bay can vary from less than 10 o/oo to more than 33 o/oo depending upon season and location. A 50-fold change in A_{Ag^+} would accompany this salinity variation while A_{AgHS^o} and $A_{AgCl_2^-}$ vary by factors of approximately 5 and 2, respectively. Both field and laboratory experiments to determine the relative bioavailability of the chloride and sulfide complexes vs. uncomplexed silver are clearly indicated. The ability of the chemical model to simultaneously compute the activity of numerous trace element complexes will facilitate the evaluation of possible synergistic toxicity effects.

The speciation portion of this type of a chemical model will also be useful in studies of the adsorption of silver or other metals onto inorganic substrates, since the first hydrolysis product is generally thought to be the adsorbing species (see review by Jenne and Luoma, in press).

We concur with most previous investigators that the activities of the $AgCl_2^-$ and $AgCl_3^{2-}$ complexes are greater than those of the other silver complexes in marine waters. However, in estuarine waters, it appears that the A_{AgHS^o} is likely to be greater than that of other silver solute species. We are unable to affirm the suggestion of Smith and Windom (1972) and Mangel (1971) that Ag_3I^{2+} and AgI_3^{2-} are the major silver species in sea water. We have not found the necessary thermodynamic data to permit us to include the former complex in our speciation and the $A_{AgI_3^{2-}}$ ranges from 7.3×10^{-22} to 3.6×10^{-20} according to our calculations. Similarly, the suggestion that Ag^+ is a quantitatively important species in marine waters (Forster et al., 1971) is not supported by our calculations.

The information most urgently needed to increase the accuracy of silver speciation is reliable analytical values for dissolved hydrogen sulfide and natural dissolved organic ligands. Thermodynamic data are also needed for the several additional inorganic complexes listed earlier in order to assess their importance. Reliable redox potentials and solubility data are required to determine the approach to saturation of Ag(s) and Ag_2S, respectively.

4.7 CONCLUSIONS

Speciation of solute silver through an estuarine salinity range using a computerized chemical model (WATEQ2) indicates that at dissolved silver and H_2S + HS^- levels of 0.000 04 and 0.000 01 mg/liter, respectively, $AgHS^o$ is an important species. In the river end member, the activity of AgHS exceeds that of Ag^+ and $AgCl^o$ by as much as 10-fold. In the marine end member, only the activity of $AgCl_2^-$ exceeds that of $AgHS^o$, and Ag^+ is a trivial quantity. $AgBr^o$, $Ag(HS)_2^-$, AgF^o, $AgOH^o$, AgI^o, $AgNO_3^o$, $Ag(NO_2)_2^-$, and $AgSO_4^-$ are all of minor or negligible importance.

These waters are oversaturated with respect to $Ag_2S(s)$ by 7 to 12 orders of magnitude, relative to available thermodynamic data. This situation is attributed to the use of thermodynamic data obtained on a highly crystalline phase, probably of large particle size, and possibly of different stoichiometry than the silver sulfide likely to precipitate in an estuary. Slow precipitation rates may also cause oversaturation with silver sulfide.

The river end member is apparently oversaturated with respect to Ag(s) if the redox potential calculated from the activities of analytically determined Fe^{3+} and Fe^{2+} is used. Use of redox potentials from platinum electrode measurements or calculated from NO_3^-/NH_4^+ or NO_3^-/NO_2^- couples indicate undersaturation of the river water with respect to Ag(s).

4.7 Appendix *Dependence of apparent activity of the various silver solute species upon the relationship used to compute the activity coefficients of neutral species.*

Solute species	Farallons		Rio Vista	
	$\gamma_{AgL} = 10^{0.1I}$	$\gamma_{AgL} = \gamma_{Na^+}$	$\gamma_{AgL} = 10^{0.1I}$	$\gamma_{AgL} = \gamma_{Na^+}$
	molal			
Ag^+	2.757E-15	2.553E-15	4.452E-11	4.359E-11
$AgBr^o$	2.993E-14	2.770E-14	5.511E-14	5.395E-14
$AgBr_2^-$	2.051E-14	1.896E-14	4.703E-18	4.213E-18
$AgBr_3^{2-}$	3.449E-16	3.187E-16	8.250E-24	8.077E-24
$AgCl^o$	2.190E-12	2.039E-12	2.752E-11	2.695E-11
$AgCl_2^-$	8.277E-11	7.747E-11	8.098E-13	7.928E-13
$AgCl_3^{2-}$	2.108E-11	1.984E-11	1.611E-16	1.578E-16
$AgCl_4^{3-}$	1.189E-11	1.125E-11	7.080E-20	6.933E-20
AgF^o	2.482E-19	2.332E-19	6.391E-16	6.257E-16
$AgHS^o$	5.180E-11	4.414E-11	2.930E-10	2.778E-10
$Ag(HS)_2^-$	2.179E-16	1.710E-16	4.320E-19	3.968E-19
AgI^o	4.023E-15	3.723E-15	2.385E-12	2.335E-12
AgI_2^-	1.773E-17	1.639E-17	3.857E-16	3.777E-16
AgI_3^{2-}	3.169E-20	2.929E-20	2.488E-20	2.436E-20
AgI_4^{3-}	5.984E-27	5.526E-27	1.754E-28	1.717E-28
$AgOH^o$	5.155E-19	4.773E-19	5.605E-15	5.487E-15
$Ag(OH)_2^-$	6.369E-23	5.898E-23	4.662E-19	4.564E-19
$AgSO_4^-$	1.439E-16	1.198E-16	4.689E-14	4.577E-14
$AgNO_3^o$	1.091E-20	1.009E-20	3.137E-16	3.071E-16
$Ag(NO_2)_2^-$	3.368E-26	1.877E-28	1.170E-21	6.902E-24

5

Biochemical Effects of Silver

D. A. Klein

5.1 SILVER-MOLECULAR INTERACTIONS

With the wide range of competing chemical species and the controlling role of halides in the formation of inorganic silver compounds, as noted in chapters 3 and 4, the concentrations of ionic silver in natural environments will be extremely low. In addition, with the information given in these chapters, it is possible to predict silver ion concentrations in specific environments. As most available information suggests that the silver ion is the primary silver form which is reactive on a biochemical level, the major emphasis of this chapter is placed in this area.

In evaluating possible silver effects on a particular organism at a biochemical level, certain dose-response relationship assumptions must be made. As discussed by Dinman (1972), the concept has been presented that there is no threshhold below which a chemical agent does not induce a biological reaction. This conclusion is based on studies in which a linear relationship was observed between carbon dioxide imposition and temporal perception changes. An assumption of a linear dose-response relationship then leaves the implication that "even one molecule will produce a biological effect, that would necessarily be deleterious to the cell." Dinman suggests that stochastic determinants, however, impose a lower limit on this relationship, and that with less than 10^4 atoms of a material per cell effects will not occur, assuming the cell to be a "concerted reaction mixture."

The question of how many molecules of a particular chemical might be required to have an effect is somewhat more complicated, since the cell is not a uniform reaction mixture, but involves surfaces and gradients. In this context, Danielli and Davies (1951) suggested that the reaction of metals, as particular toxic components, is most critical at interfaces, and especially at the cell surface. Depending on the charge of the particular metal, it may combine with ionic groups on the surface by both electrostatic and covalent bonding. Much of this earlier work has shown a correlation between electrode potential and metal toxic properties (Danielli, 1944). However, a more realistic way to measure metal interactions with

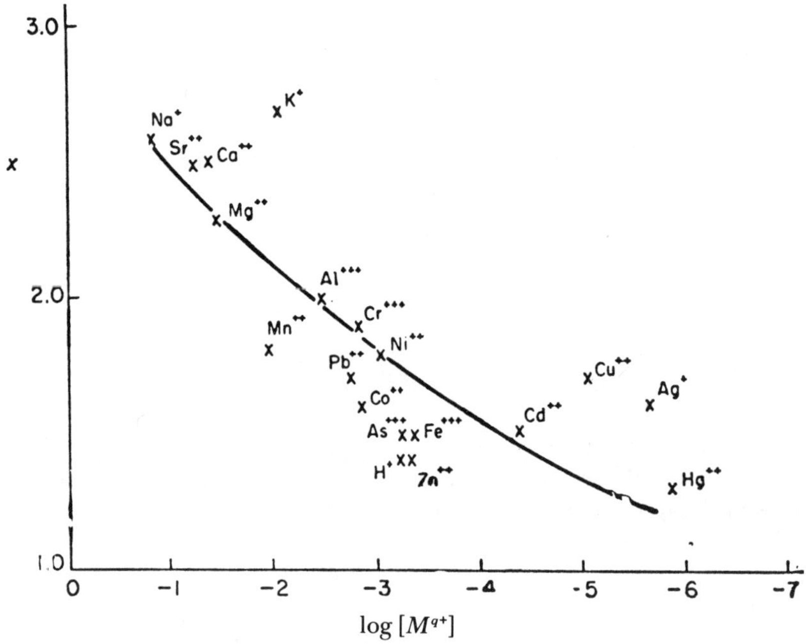

FIGURE 5-1. *Plot of electronegativity differences between metal and oxygen atoms, x, against the logarithm of the toxic concentration of metal ions (Danielli and Davies, 1951).*

cell surfaces involves measuring differences in electronegativity between the particular ion and oxygen, as noted in Fig. 5-1. Using this approach, silver and copper appear to be toxic at lower concentrations than would be predicted, as a result of their high energy of hydration once they have been detached from the surface. What is of interest is that copper, mercury, and silver are most toxic at the lowest concentrations, being less than 10^{-5} M.

Eichhorn (1975) has provided an excellent summary of possible ways in which heavy metals may interact with active sites of macromolecules. These active sites can involve metals, which may be displaced by other metals; in addition, every electron donor group on the molecule can potentially bind metal ions. These include a series of functional groups which can bind or otherwise interact with metals, including the amino, carboxyl, guanidino, hydroxyl, imidazole, peptide bond, and sulfhydryl groups.

Toxic metals can interact not only with the active site, but by other types of interactions as well (Eichhorn et al., 1969). In addition, the added metals may not bind at the site where the original metal was connected to the enzyme, which implies that the same metal may have beneficial or adverse effects on an enzyme at some particular concentrations (Fig. 5-2).

Most metals also show activation at lower concentrations by binding active sites, and at higher concentrations when they bind to inhibition-related sites.

In summary, Eichhorn noted the following ways in which metals can interact with macromolecules:

1. as a requirement in subunit aggregation,
2. by interference with hydrogen bonding (e.g., heavy metals can bind to bases competing with hydrogen bonding, resulting in destruction of double helix structures),
3. through competition with magnesium binding to phosphate groups,
4. by influencing the winding and unwinding of DNA when heated, making this a reversible process,
5. by leading to mispairing of nucleotide bases, resulting in mistakes in protein synthesis with the wrong metal concentration present, and
6. by depolymerizing biomacromolecules by binding to phosphate groups and making phosphodiester linkages more susceptible to hydrolysis.

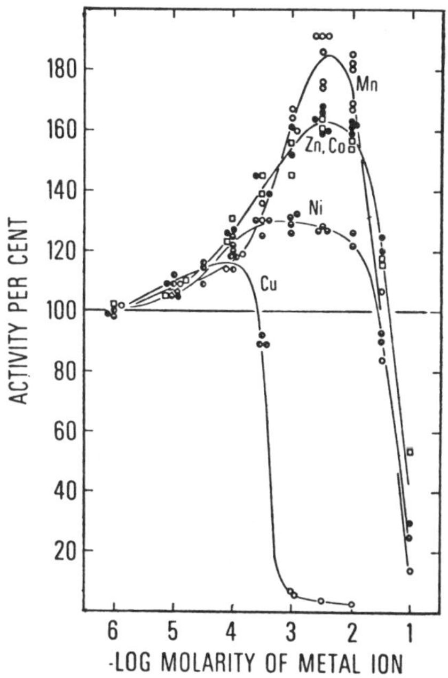

FIGURE 5-2. *Effect of metal ions on the enzymatic activity of ribonuclease (Eichhorn et al., 1969).*

Gillen et al. (1964) completed an excellent study on silver ion binding to adenine and substituted adenines which emphasizes many of the points brought out by Eichhorn. These workers found that adenine and 6-dimethyl aminopurine, both of which have N-9 hydrogen atoms, were able to bind silver (with hydrogen ion displacement at silver ion concentrations approaching 10^{-8} M (Fig. 5-3). Other adenine derivatives which were blocked at the N-9 hydrogen atom, but which had amino hydrogens, bound much less strongly at concentrations above 10^{-4} M. The silver ion amounts required to be in solution for a certain degree of binding were also pH-dependent (Fig. 5-4). At lower pH values, a higher silver ion concentration had to be in solution to have an equivalent amount of binding. Similar pH dependencies were observed for different levels of silver binding.

Daune et al. (1966) carried out detailed studies on silver ion interactions with natural and synthetic polynucleotides where silver complexes were formed which complement the previous study. Using a variety of analytical techniques, these workers found a strong temperature-sensitive reversible complex between silver ions and nitrogen atoms. Silver ion bound to four bases, provided two of them were a G-C pair. A second temperature-insensitive complex could be formed with proton release. In addition, silver presence increased tobacco mosaic virus RNA resistance to temperature changes and ribonuclease presence.

To investigate metal interactions with biomacromolecules, radioactive metal forms have also been used (Sabbioni and Girardi, 1977). These methodological advances (neutron activation analysis, multiple tracing, high resolution x-ray spectrophotometry and Cerenkov counting), when coupled with standard biochemical techniques, allow a much more precise study of several unique and poorly understood aspects of metallobiochemistry.

5.2 EFFECTS ON ENZYMATIC REACTIONS

Much information is available on specific effects of silver (chiefly in the ionic form) on endergonic and exergonic enzymatic reactions. The more conventional studies have involved hydrolytic (exergonic) enzymes of various types, as discussed by Smith and Carson (1977). Berezin et al. (1971) in a study of enzyme active center flexibility, observed lead and silver ion inhibition of α-chymotrypsin. Based on kinetic data, the binding sites for each of these metals were different, and the binding of one of these metals facilitated the binding of the second ten-fold.

An additional technique widely used to detect minor changes in enzymes due to the presence of silver ion is luminescence spectroscopy. Chen (1975) found that silver ion binds at the amino group of tryptophan,

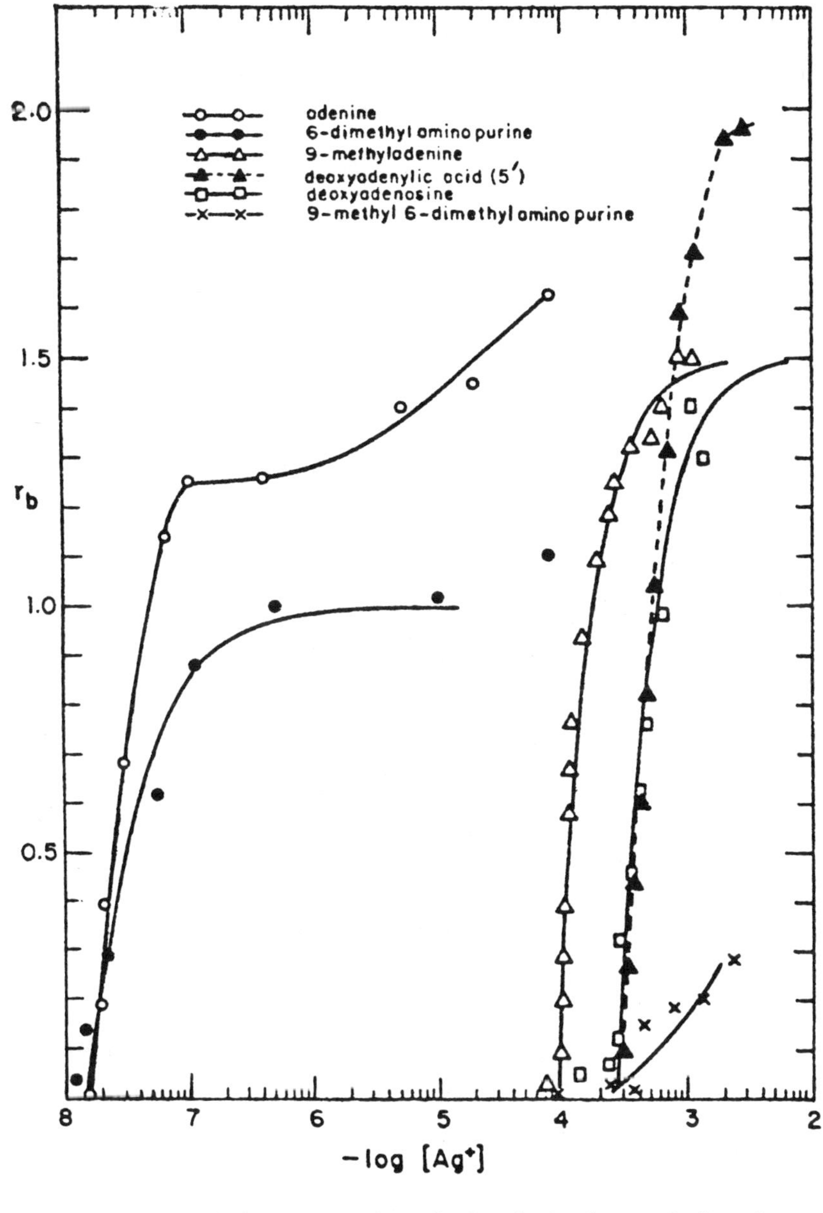

FIGURE 5-3. *Binding curves for adenine derivatives and silver ion at pH 7 in 0.1 M $NaClO_4$ at 25 ± 1°C. The initial purine concentrations is always 10^{-3} M. The quantity r_b is defined as the moles of silver bound per mole of purine (Gillen et al., 1964).*

and that quenching is a complex process involving ultraviolet absorption changes due to mercaptide bond presence. Rogers (1969) used fluorescent techniques and noted that sulfhydryl-containing dehydrogenase enzymes did not react solely by the formation of mercaptide bonds, and also reacted

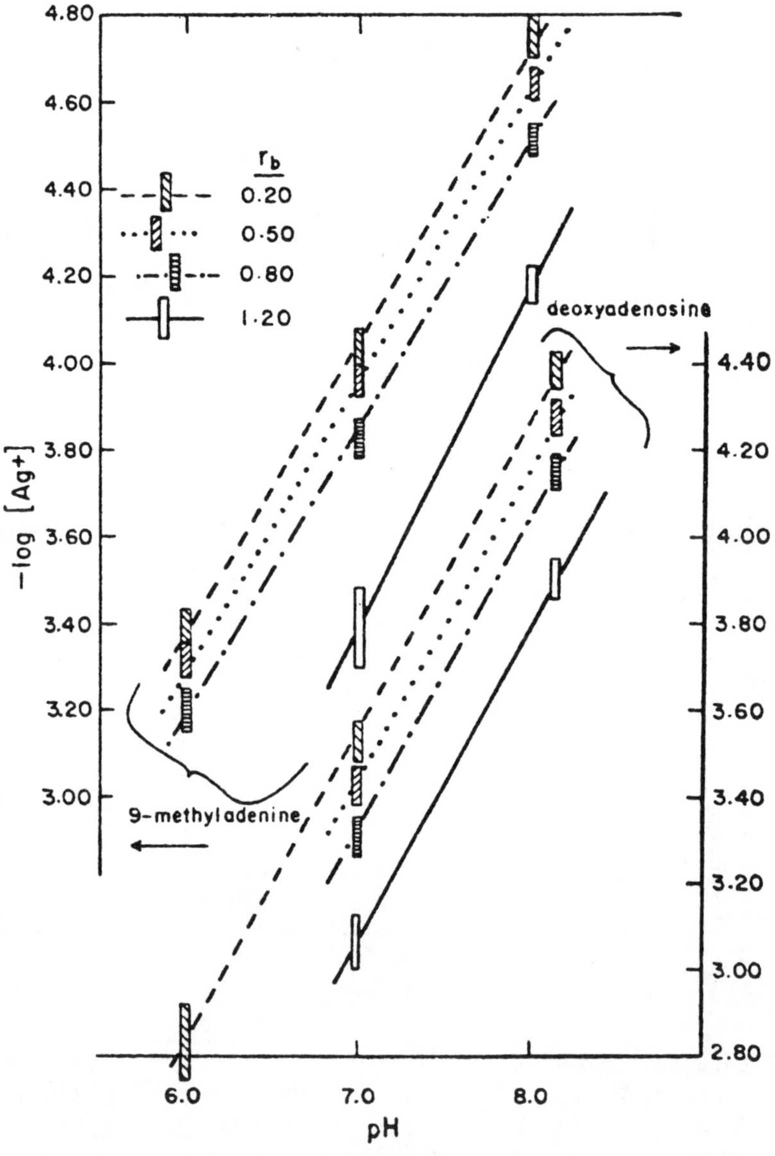

FIGURE 5-4. *Variation of log Ag^+ with pH at several r_b values for deoxyadenosine and 9-methyladenine (Gillen et al., 1964).*

by relatively displacing tryptophan residues from a nonpolar to a polar environment, and by the inability to bind $NADH_2$. These changes were quite rapid, taking only 10 minutes, while 3 hours were required for precipitation of denatured protein. Nonsulfhydryl group-containing proteins (α-chymotrypsin, ribonuclease, bovine heart cytochrome C) did not show altered fluorescence spectra. The importance of sulfhydryl groups in protein reactions with silver (and other metals) was also emphasized in work carried out by Montie and Montie (1973), in which plague murine toxin was detoxified by an approximate 1:1 ratio of silver ion to sulfhydryl groups; and in studies by Guilloux et al. (1968) on May Bug trehalase.

The sulfhydryl groups are not the only active sites. Work by Euler and Svanberg (1920), and later studies by Myrback and Willstaedt (1957) and Myrback (1963) have shown that yeast invertase could be inhibited noncompetitively by silver ions, and that this reversible inhibition can occur by the addition of compounds forming stable silver complexes, or by dialysis. In comparison with sulfhydryl-containing compounds (cysteine, glutathione, thiogalactic acid) where the complexion of the silver by the enzyme is strongly dependent on pH, that of nonsulfur containing substances is not.

Schmitt et al. (1975) studied the inhibition of subunit A of lactic dehydrogenase by the sulfhydryl reagent $AgNO_3$, and found that subunit A (containing SH groups) was inhibited, while the non-SH-group containing subunit B was not. Of interest was the observation that the lactate dehydrogenase isoenzymes which function predominantly under anaerobic conditions contain the subunit A and are sensitive to silver ion presence (LDH 5), whereas the aerobic form (LDH 1), representative of tissues with aerobic metabolism, was not sensitive to the presence of silver.

Several other studies on the possible role of SH groups in the sensitivity of enzymes to the presence of silver are of interest. Snodgrass et al. (1960) found that the activity of a yeast alcohol dehydrogenase was dependent on the presence of zinc and -SH groups, and that silver ion could instantaneously inhibit the activity of this "sulfhydryl" enzyme. This inhibition was entirely independent of the presence and functioning of the zinc atoms.

Higashida et al. (1975) noted that mitochondrial malate dehydrogenase was influenced by silver ion, both by effects on structural regularity and attacks on sulfhydryl groups. In a similar study, Gorin and Chin (1965) observed that native urease was inhibited 50% by eight silver ions per enzyme molecule. However, if the 21 reactive mercapto groups were previously combined with N-ethylmaleimide, with eight silver ions present per enzyme molecule, 100% inhibition occurred. Kaneshiro et al. (1975) modified β-galactosidase with silver ion to form inactive monomers, which could be recombined with an excess of thiol to reform active dimers and tetramers, showing the reversibility of this type of process.

Some evidence is also available indicating that histidine moities of enzymes can interact with silver ion, leading to changes in enzymatic activity. Martinek et al. (1969) found that chymotrypsin can be reversibly inactivated by silver ion which binds with histidine imidazole groups, and Hough and Shall (1970) showed a similar pH-dependent sensitive site for a phosphate-free-N-acteyl homocysteinyl-ribonuclease. The original rhibonuclease A was not affected by silver ions.

Silver also interacts with coenzyme-mediated enzymes by interference with flavin adenine nucleotides (FAD). Nakamura and Ogura (1968) found that a non-SH-group containing glucose oxidase-FAD enzymatic system was sensitive to the presence of silver ion but not to nonmetallic SH-reagents. It was concluded that silver ion inhibited the oxidation of the reduced FAD moiety, competing with molecular oxygen used as a hydrogen (electron) acceptor. The kinetic data also indicated that the binding sites for silver and mercury were different.

In more complex systems which involve associated coenzymes and cofactors, the specific sites for silver ion effects can become more complex. Kratochvil et al. (1967) studied the effects of a series of metals on isocitric dehydrogenase (which catalyzes the conversion of isocitrate to α-ketoglutarate in the presence of NADPH and a metal, usually manganese, magnesium, or cobalt) while developing new methods for metal determination. The metal in this enzyme system was felt to be directly involved in the binding of isocitrate to the enzyme, which could occur at two sites. Silver ion was found to cause a 35% inhibition of this enzyme at 3.1×10^{-8} M.

The effects of silver ion on energy-yielding processes in suborganelles have also been evaluated. In an excellent general summary of the effects on active transport processes, Harold (1974) emphasized the chemiosmotic interpretation of energy production as related to vectorial metabolism, where a membrane potential can yield energy. Rayman et al. (1972) found that silver ion was a very potent inhibitor of succinate uptake and energy coupling in membrane vesicles prepared from *Escherichia coli*. This effect was assumed to be due to silver ions functioning as a sulfhydryl group reagent, and at 10^{-8} M, distinct inhibition was observed.

In the related area of cyclic adenosine monophosphate (c-AMP) metabolism, Nathanson and Bloom (1976) studied ionic metal effects on adenylate cyclase and cyclic AMP phosphodiesterase, the enzymes responsible for cyclic AMP formation and breakdown. At an eight micromolar concentration, silver ion caused a 50% inhibition of adenylate cyclase, and using a series of metals at 0.1 micromolar, silver ion had the greatest affect on cyclic AMP hydrolysis. These workers felt that the concentrations used in this study approached those found following toxic exposure to certain metals, as discussed by Vallee and Ulmer (1972), and indicated that metal ions at "very low concentrations" could have marked effects on cyclic AMP metabolism *in vitro*.

In mitochondrial systems, the effects of silver ion on adenosine triphosphate were found to be generally stimulatory, with or without the passage of an electrical current (Chappell and Greville, 1954). Of interest was the observation that dinitrophenol (an electron transport chain inhibitor) also caused a similar effect. By use of 10^{-5} M silver ion, the ATP activity of mitochondria was increased five-fold. This was related to a doubling of the respiratory rate using pyruvate and malate as substrates. This effect has been generally related to the role of ATP, in excess, in inhibiting respiration, perhaps as an allosteric control component. In a similar study, Primack (1975) found that the presence of gold, copper, or silver produced mitrochondrial swelling and increased mitochondrial protein synthesis. These studies suggested, at least for copper, that this metal may have a role in mitochondrial protein synthesis regulation, since the levels were compatible with those found *in vivo*.

In contrast with mitochondria where respiration processes occur, the lysosomes (suborganelles where hydrolytic enzymes such as ribonuclease and acid phosphatase are segregated) of higher cells can protect cellular homeostasis through the sequestration and chelation of metals. As noted by Sternlieb and Goldfischer (1968), this process has usually been associated with argyria. In the liver an association between silver and the lysosomes of hepatocytes and Kupffer cells was noted. Similar studies have been carried out by Misaka and Tappel (1971) using rat liver lysosomes.

The effects of silver ion on the RNA polymerase activity of isolated rat liver nuclei were studied by Novello and Stirpe (1969) where silver ion, together with mercury, cadmium, and copper at millimolar concentrations, strongly inhibited activity. Similar studies of silver inhibitory effects on plasmin activity have been carried out by Kowalski et al. (1957). In addition, available literature on silver effects on copper metabolism have been summarized by Evans et al. (1970).

On a physiological and biochemical level, the basis for metal tolerance in plants and microorganisms has been considered by Ernst (1976). Microorganism tolerance to metals was felt to be achieved by high membrane affinity for heavy metals, or by the evolution of metal-resistant enzymes. Generally the sulfhydryl group-containing enzymes are very sensitive to heavy metals, and the functioning of such enzymes may be due to the ability of the microorganism to maintain a pool of sulfhydryl groups, or to form stable metalloprotein compounds.

5.3 CONCLUSIONS

The biochemical reactions of metals appear to be most important at interfaces, such as the cell surface. It has been suggested that observable changes will not occur unless more than 10^4 atoms per cell are present, if the cell is assumed to be a "concerted reaction mixture."

Toxic metal interactions may include interference with hydrogen bonding, competition with essential trace metals, influences on DNA winding and unwinding, effects on subunit aggregation and depolymerization of macromolecules. These interactions will be dependent upon pH, temperature, concentration and form of silver, and the ability of the cell to deactivate silver (tolerance). Studies of metal-biomacromolecule interactions have been greatly aided by use of radiotracer techniques.

Silver may react with both sulfhydryl-containing enzymes and those without sulfhydryl groups. Active transport across membranes, and cyclic-AMP metabolism may also be affected, even though sulfhydryl groups may not be present.

Low levels of silver (10^{-5} M) can stimulate mitochondrial ATP activity, resulting in increased respiration. Gold, copper, and silver have also been linked to increased mitochondrial protein synthesis, but only at extremely low concentrations. Toxic effects are generally observed with silver concentrations higher than approximately 10^{-4} M.

Low silver ion concentrations can be expected to be present in natural environments, due to interacting chemical species and the presence of absorptive organic materials and clays. Thus, after chemical and physical equilibration of silver with the environment, effects of silver ions may be difficult to detect, and it is possible that stimulatory as well as inhibitory interactions might occur.

6

Effects on Microorganisms

D. A. Klein

6.1 INTRODUCTION

Information on environmental exposure and potential environmental impacts of silver-containing nucleating agents is given in the reviews of Cooper and Jolly (1969, 1970), the annotated bibliography on biological and physiological effects of silver iodide used in cloud seeding by Grimes (1972), the general review on silver exposure in the environment by Smith and Carson (1977), and the summary of environmental impacts of precipitation management, based on Project Skywater (Howell, 1977).

Generally, microorganisms are considered to be more sensitive to the presence of silver than higher organisms, both plant and animal, and the well established antimicrobial activity of silver has been used widely in medicine and sanitation.

This known toxicity of silver ion to microorganisms also has been cited to stress the potential toxicity of silver iodide-derived nucleating agents, although direct toxicity of silver iodide at environmentally-relevant concentrations is difficult to demonstrate. As low but predictable levels of silver ion can be released from silver iodide due to ionization processes, it is important to consider the potential effects of both silver ion and silver iodide on microorganisms and microbial activities.

6.2 EFFECTS OF PHYSICAL PARAMETERS ON SILVER ANTIMICROBIAL ACTIVITY

Much of the earlier work on the effects of physical conditions on the activity of silver ion, derived either from soluble silver compounds or metallic silver, was related to the use of silver in water purification and to the concept of "oligodynamie." This term concerns the effect of a compound at very low concentrations; and in the case of silver ion, both stimulation and inhibition have been observed, depending on the test organism and the particular conditions.

Excellent summaries on the historical aspects of silver use as a germicidal agent are given by Just and Szniolis (1936) and Porter (1946). Early workers noted that the group I elements copper, silver, and gold were especially effective as germicides at low concentrations (Seuderling, 1933). When the pure metals were placed in contact with microorganisms on a solid growth medium, the following conclusions about silver effects on microbes were drawn:

1. Organisms not growing near the silver were not killed, but could initiate growth in fresh silver-free culture media after a lag period.
2. Death occurred more rapidly with higher free silver concentrations.
3. Proteins (as from agar) protected microbes from the effects of silver ions.
4. Organisms which had been exposed to this metal could possibly lose their pigment production capabilities.

As noted by Just and Szniolis (1936), Nageli (1893) observed that the alga *Spirogyra* would not survive in water which had a 10 μg/liter silver ion concentration derived from sodium nitrate. Even using the more sensitive procedures presently available, the threshhold level for observing effects has been decreased, at most, by two log units in over almost a century of work.

The earlier literature on the "oligodynamic effects" of silver is voluminous, and has been summarized by many recent reviewers. In spite of this, there are some additional points from the older literature which warrant discussion.

Doerr (1920a,b; 1921) and Doerr and Berger (1922) completed a series of extensive studies on silver oligodynamic effects. Working primarily with pathogenic microorganisms, the questions of buffer effects on the activity of silver ion and on microbial resistance development were raised. Concerning the latter, he suggested that water bacteria were not inherently resistant to this metal. He also noted that silver ion would stimulate microbial growth at the lowest concentration used (Table 6-1). In the second of these studies, Doerr (1920b) further examined the stimulation of bacterial growth on agar plates and noted increased growth immediately adjacent to the inhibition zone, a phenomenon which has been discussed extensively since that time. In addition, he noted that flaming or boiling of metallic silver would eliminate the antimicrobial activity, which could be restored by contact with strongly dissociated acids or oxygen. In this paper he also noted that:

1. the active principle is dialyzable,
2. with different silver forms, similar zone-causing effects were observed,

TABLE 6-1. *Effects of silver ion concentration, contact time and sodium chloride presence on the survival of a test culture, expressed in microorganisms/ml (Doerr, 1920a).*

Elapsed Time	Control Culture Population	$AgNO_3$ Concentration and Diluent							
		1:500		1:500,000		1:5,000,000		1:5,000,000,000	
		in Aq.	in NaCl	in Aq.	in NaCl	in Aq.	in NaCl	in Aq.	in NaCl
30 min.	16,000	0	2,000	0	3,000	0	16,000	16,000	8,000
4 hr.	17,000	0	0	0	0	0	500	16,000	20,000
9 hr.	18,500	0	0	0	0	0	20	32,000	45,000

(Doerr, 1920)

3. increased colony size, and not increased numbers of colonies were observed adjacent to the zone of inhibition, and
4. in defibrinated blood agar plates, silver can cause hemolysis.

Doerr (1921), in the third of his extensive studies, formulated additional concepts which directly apply to the analysis of possible impacts of silver-based nucleating agents in the environment. He noted that after flaming, the bactericidal activity could be recovered in free air but not in liquid paraffin, thus implicating atmospheric components in this process. In the last of these studies Doerr and Berger (1922) considered the problem of whether silver ion was the only cytotoxic factor, or whether molecular or particulate silver might have effects. By careful neutralization experiments using cyanide salts, they concluded that silver ion was the active component, and that CO_2 and O_2 could cause the formation of an active silver compound. Similar results were obtained by Buschke et al. (1925) in a study where the then recently completed work of Doerr was not cited.

In the same vein, Hosenfeld (1938a) discussed the oligodynamic activity of silver, repeating many studies already available in the literature. He observed the ability of extremely low silver ion levels to stimulate microorganisms and discussed possible mechanisms of silver antimicrobial activity (internal metabolism and sulfhydryl group effects). In addition, he provided information on how environmental conditions affect the activity of silver ion against microbes. He noted that at higher temperatures microbial die-out was faster, that impurities retarded antimicrobial activity, and that bacteria could become accommodated to the presence of silver (to a level of 50 mg/liter by a "*Bakterium xylinium*"). Unfortunately, the specific experimental conditions he used were not described. A second paper (Hosenfeld, 1938b) discussed applications of silver for water supply disinfection.

Much interesting information on the antimicrobial activity of silver became available with studies by Zimmermann (1952), which repeated many earlier experiments. How to determine the mechanism of the association between silver ion and a bacteria cell (polar absorption versus phase distribution and/or chemical binding), and procedures to differentiate between these mechanisms were discussed. He noted that the positive silver ion will be attracted to the negatively charged microbial cell wall which can lead to a neutralization of charges on the cell, and emphasized that this polar absorption is reversible, since bacteria can recover from the inhibiting effect of silver ion. The earlier literature, as cited by Zimmermann, also suggests that silver ion will be equally effective under aerobic and anaerobic conditions, while metallic and colloidal silver will be less effective under anaerobic conditions.

Of major importance to more recent studies, Zimmermann cited workers (Bertuzzi, 1936) who had shown that silver-resistant bacterial strains frequently showed a darkened colony, and noted that these organisms could cause silver reduction. The original report of metal reduction, as cited by Zimmermann (1952), was made by Zsigmondy (1898) who used fungi.

More recent information on the oligodynamic activity of silver has been provided by Romans (1968). Copper ions were found to enhance the antibacterial effectiveness of silver ion, and a formulation of silver, copper, zinc nitrate and dextrose (as a reducing agent) was described as a bactericidal agent. Based on available information, Romans concluded that oligodynamic forms of metals are available which are nontoxic to man.

One of the better and more widely quoted recent papers on the role of temperature and pH in silver-microbe interactions was completed by Wuhrmann and Zobrist (1958). In this study they washed their organism (*E. coli*) carefully before use in the silver interaction system, and they also assured that the nutrient agar medium would neutralize any active silver ion which might have been carried over from exposure to the enumeration step of the study. Their conclusions can be summarized as follows:

1. An increase of the water temperature by 10°C increased the *E. coli* die-out rate 1.6 times, independent of the pH value and the silver concentration.
2. By decreasing the hydrogen ion concentration 10-fold (increasing the pH by 1 unit), silver ion effects on the test microbial populations increased 1.55-fold, independent of the temperature and the silver ion concentration.
3. With increased amounts of chloride ion, the die-out rate decreased, and the silver ion concentration decreased asymtotically.
4. Decreased silver antimicrobial activity without oxygen present was observed, although other factors were felt to contribute (that had not been fully elucidated).

In addition, Wuhrmann and Zobrist carried out experiments to establish the relative sensitivity of a range of soil, water, and medically important organisms to the presence of silver ion under standard conditions. Pathogens which caused typical water-borne diseases (*Salmonella*, etc.) were found to be at least as sensitive to silver as *E. coli*. Lacking closely controlled test conditions (pH value, temperature, salts, etc.) they did not feel that they could compare their results with those in the previous literature. Bruni (1942), using a yeast, noted that metallic silver stimulated CO_2 evolution under fermentative conditions, which complicates the assessment of silver effects under aerobic versus anaerobic conditions. In a related earlier study, Bruni (1938) found that UV light could activate silver's antimicrobial activity.

The papers by Wuhrmann and Zobrist (1958) and Zimmermann (1952) were briefly reviewed by Woodward (1963), who also discussed the tendency of silver to adhere to surfaces in relation to the inactivation of silver. In a related experiment-based paper, Chambers et al. (1962) evaluated substances which could be used to inactivate or neutralize silver ion in the enumeration phase of an experiment. For best results, a neutralizer with 500 μg/liter sodium thioglycollate and 730 μg/liter sodium thiosulfate, using 1.0 ml of sample and 9.0 ml of neutralizer, was recommended. In addition, these workers found that boiling glassware in sodium chloride solutions was an effective cleaning procedure. In the laboratory, these neutralization methods have proven themselves satisfactory under a wide range of experimental conditions.

All the studies discussed so far involved the use of pregrown microbial populations. In studies with *Hyphomicrobium* and *E. coli*, Klein and Giangiordano (1976) found that if growth instead of die-out is monitored, silver ion can cause increased growth delays at lower temperatures (Fig. 6-1). Especially interesting was the increased minimum growth temperature with silver ion present. If these cells inhibited by silver presence at low temperatures were transferred to a more suitable temperature, and the silver ion was neutralized, survivors were able to grow normally. No such effects

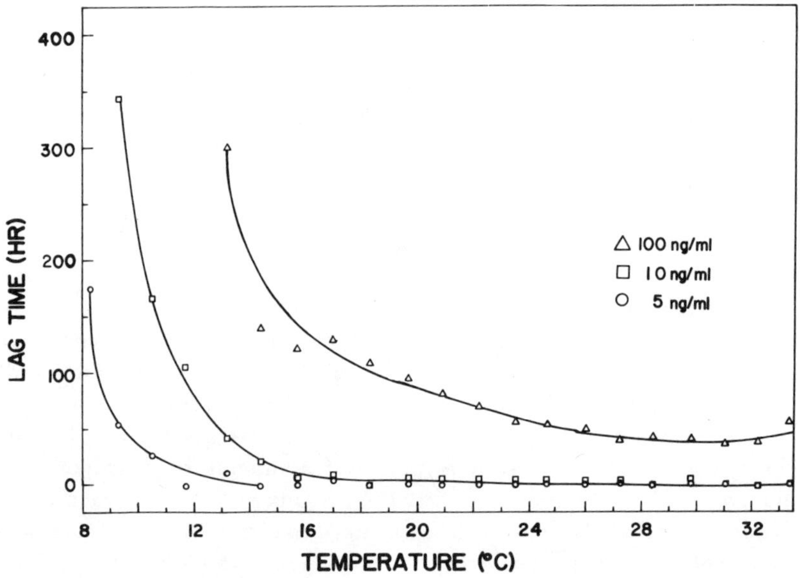

FIGURE 6-1. *Temperature effects on growth initiation by* Hyphomicrobium *in the presence of three levels of silver ion (Klein and Giangiordano, 1976).*

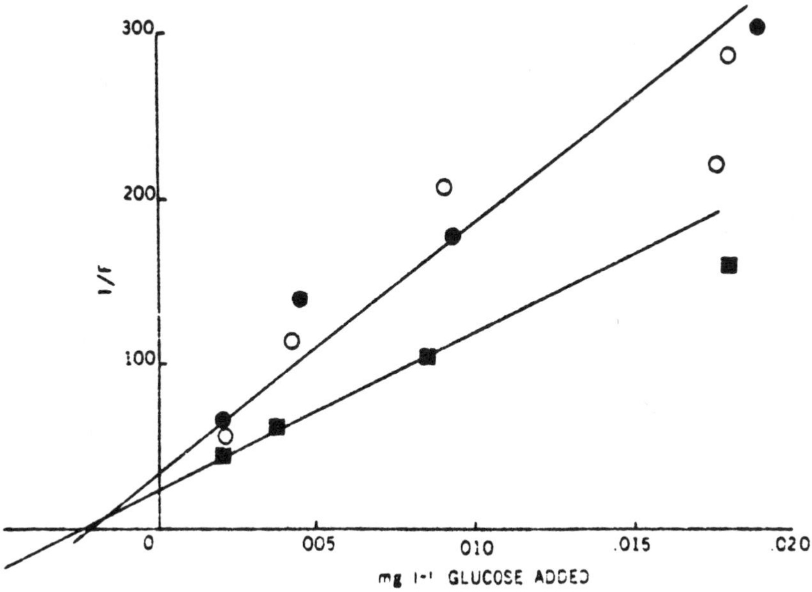

FIGURE 6-2. *Modified Michaelis-Menten plots of ■, albumin (0.50 μg/liter); ●, Ag⁺ (0.05 μg/liter); and ○, albumin (0.50 μg/liter) + Ag⁺ (0.05 μg/liter) treated water on glucose mineralization by a mixed aquatic microbial population (Albright and Wilson, 1974).*

on growth initiation were observed when silver iodide burn complexes were used.

In studies by Albright et al. (1972), and Albright and Wilson (1974), silver ion at 1 μg/liter caused decreases in microbial populations in natural water, and at these low levels cystine additions were not able to neutralize the silver effects. Using silver ion at 0.05 μg/liter, changes in glucose dissimilation were noted, with or without albumin present (Fig. 6-2). In spite of these changes in metabolic activity, the viable counts remained constant.

In recent years, many repetitive studies on silver ion and other metal ion effects on various microorganisms have been carried out by Russian workers, including studies by Avakyan (1967), Golubovich (1974), and Golubovich and Rabotnova (1974). Other than using a different series of organisms, no new concepts have been provided, except perhaps the observation of greater resistance to silver by yeasts than by the bacteria.

An interesting paper published by Annear and Mee (1976) noted a "paradoxical" response of bacteria to silver nitrate, where greater effects were seen at lower rather than higher concentrations. The silver-sensitive variants of five bacterial genera showed inhibition from 5.7 to 18 μg/ml of silver nitrate while at higher concentrations (24, 32, 42 μg/ml) growth was observed.

Although a wide variety of sanitizing agents have gained greater acceptance than silver, research on the antimicrobial activity of silver continues unabated. Brown and Anderson (1968) studied the effects of silver ion on *Pseudomonas aeruginosa*, and noted that "little information is available." Using cells grown at pH 6.25 with glucose-magnesium or magnesium-limited cells, increased die-off was observed, compared to glucose-limited cells (Fig. 6-3). This indicated that the cell walls of magnesium-limited cells provided a less effective barrier to absorption and entry of silver ion into the cells. Since decreased silver effectiveness was noted at lower pH values, these workers suggested that there was greater competition between the carbon and hydrogen ions for anionic sites on the bacteria.

6.3 COMMERCIAL AND INDUSTRIAL APPLICATIONS

Research on the commercial and industrial applications of silver as a bactericide is also continuing. Silver metal use as an antimicrobial agent in home water filter units containing activated carbon (Stern, 1974), and applications as a microbiocide in hydrocarbon fuel compositions (Baptist et al., 1965) have been described. With the need to maintain a potable water supply on space vehicles, silver has been used as a disinfectant in USSR programs (anonymous, 1974), where silver at 0.1 and 0.2 μg/liter did not accumulate, and has been reported to have no effects on the functioning of organs of experimental animals. In addition, year-long tests with human subjects gave satisfactory results. A major problem was the selection of a container material which would minimize silver adsorption and inactivation (plastics were recommended).

Similar studies have been carried out by Jasionowski and Allen (1974) related to American space research. Tests under simulated orbiter conditions showed that silver ion at 0.05 μg/liter was bactericidal against *Pseudomonas aeruginosa*, and design specifications for a silver ion generator were provided which could be used under spaceship conditions.

Another impact of silver (at least as silver ion) in the environment is on commercial copper extraction from mine tailings, using *Thiobacillus ferroxidans*. Both Imai et al. (1975) and Hoffman and Hendrix (1976) have shown that silver ion can inhibit ferrous ion oxidation, which could decrease copper recovery. Regarding the opposite effects of silver, the British Columbia Research Council in a recent patent (McElroy and Duncan, 1974), developed a procedure which uses silver ion to enhance copper recovery from leached ores. This is important if the presence of nu-

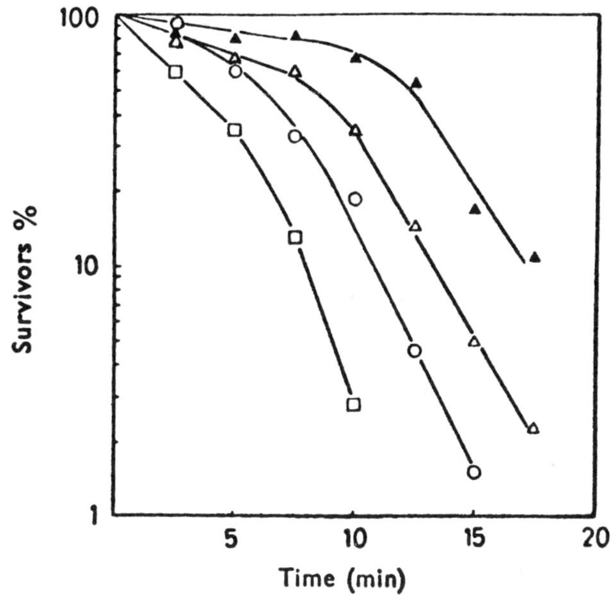

FIGURE 6-3. *Rate of kill of* Pseudomonas aeruginosa *in 1 × 10^{-5} M silver nitrate at 25°.* ▲ *Glucose-limited cells at pH 5.85;* △ *glucose-limited cells at pH 6.25;* ○ *glucose-magnesium limited cells at pH 6.25;* □ *magnesium-limited cells at pH 6.25 (Brown and Anderson, 1968).*

cleating agent-derived silver leads to increased silver in waters used for mineral leaching.

6.4 EFFECTS ON VIRUSES, FUNGI, YEASTS, AND SPORES OF BACTERIA AND FUNGI

The effects of silver on organisms other than typical procaryotes, including various viruses, filamentous fungi and spores (bacterial and fungal) have been investigated. Generally, the spores of most bacterial species have been found to be more resistant to silver than were related vegetative organisms (Kondratyev, 1971). Cliver et al. (1971), as noted by Jasionowski and Allen (1974), found that *Bacillus cereus* spores were resistant to 0.24 μg/g silver ion, whereas the vegetative cells of the same organism were extremely sensitive to silver ion. The ability of silver ions to modify the radiation sensitivity of bacterial spores is also of interest. Richmond and Powers (1974), using spores of *Bacillus megaterium*, found that silver ion effects on the radiation sensitivity of the spores depended on the presence or absence of oxygen. Without oxygen, silver ion sensitized the spores to xrays in a concentration-dependent fashion. With oxygen, the silver caused an "over CO_2" effect, dependent on the O_2 concentration. The competition between O_2 and silver ion for electrons which would normally transform the OH radical to a harmless hydroxyl ion was felt to cause this effect, which increased the radiation resistance of the spores. The optimal silver concentration, with or without oxygen, was 2×10^{-3} M.

White and Munns (1951) found that silver (together with cadmium, copper, osmium, mercury, and palladium) was extremely toxic to yeasts, reducing growth under fermentative conditions at 0.22 μg/g, and that this effect was diminished in the presence of malt extract or molasses, as would be expected.

Fungal spores also are sensitive to silver ions, and this sensitivity can be influenced by the presence of other ions, especially the halides. Miller and McCallan (1957) observed that rapid silver uptake by fungal spores, with contact times of 1 minute or less, led to germination inhibition. This toxicity was not affected by chloride ion, but it was reduced by bromide and prevented by iodide. Silver ion had a marked effect on spore permeability as measured by radioactive phosphorus release, with mercury and copper ions being the most effective of the metals tested. Somers (1959) tested the sensitivity of spores to silver ion in an attempt to test the postulation of Danielli and Davies (1951) relating electronegativity and toxicity. Using conidia of *Alternaria tenuis*, the linear relationship proposed by Danielli was achieved (Fig. 6-4). Somers felt that the toxic action of metal cations involved the formation of a nonionized complex with surface ionogenic groups such as phosphate, carboxyl, or sulfhydryl groups; and that the different metal toxicities could be related to various surface binding strengths.

Nielsen and Massey (1940) expressed hope for silver effectiveness as a fungicide, based on the relative sensitivity of pathogenic fungi to this metal. *Aspergillus niger* grew feebly in a medium held in silver dishes, with the retardation being attributed to silver ions released. Another experiment showed delayed *A. niger* growth at a silver concentration one-third of that required to produce the same effects with mercury. Silver iodide gave a 58.7% inhibition of fungal spore germination, while numerous other silver compounds inhibited germination up to 100%. Silver was effective as a seed disinfectant, however, silver nitrate could destroy the embryos of wheat, pea, and barley seeds with broken seed coats. They also cited an experiment using 10 and 100 μg/g silver as a soil disinfectant, in which seedlings were so weakened by the high added silver levels that intensified fungal growth occurred.

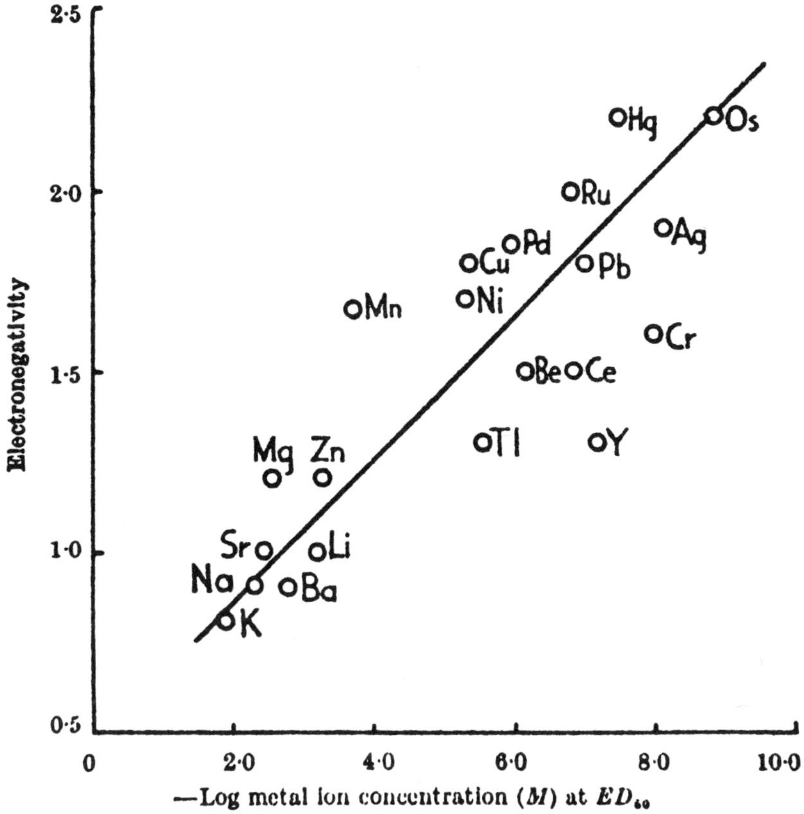

FIGURE 6-4. *The effective toxic dose (ED_{50}) of metal cations to* A. tenuis *as related to their electronegativity (Somers, 1959).*

There is also information available on the possible effects of silver ion on various types of viruses. Jasionowski and Allen (1974), citing various sources, indicated that silver ion in the range of 0.05 to 0.25 μg/g inactivated rhinovirus type 1A, reovirus 1, several enteroviruses, and influenza virus A. Vaccinia, representing the pox virus group, appeared to be completely resistant to silver ions, which may be a reflection of the more complex wall structure of these viruses. Shimizu et al. (1976) found that herpes simplex virus types 1 and 2 were inactivated by silver nitrate at 30 μM or less, a level which did not affect the hemagglutination characteristics of five other plant and animal viruses. The inactivated virus could be absorbed to the cells and induced specific agglutination reactions although the virus particles could not be replicated.

6.5 MECHANISMS OF SILVER EFFECTS ON MICROORGANISMS

With the extensive information that is available on the effects of silver (as the silver ion) on biochemical processes, it is interesting to consider the whole microbial cell, to determine if there might be additional ramifications for the evaluation of nucleating agent impacts in natural environments.

Although the major site of silver interactions with living systems is with sulfhydryl groups, a wide range of additional functions can be affected (causing both stimulation and inhibition). The major sites of possible silver ion action, as reviewed by Spadaro and Becker (1976), are the respiratory chain, the binding of silver ions to cellular DNA, and by binding to the cell membrane. Their study suggested a silver ion effect on the expression of β-galactosidase by interactions at the cell membrane.

Bragg and Rainnie (1974) found that silver ion inhibited the oxidation of a series of carbohydrate substrates in resting *E. coli* suspensions, and that the site most sensitive to inhibitors was between the b-cytochromes and cytochrome A_2. The second most sensitive level of inhibition was at the NADH and succinate dehydrogenase respiration sites situated on the substrate side of the flavin components. In most of these studies, silver was used at 86 μM concentrations.

Extremely interesting work on the interactions of silver ion with DNA of *Pseudomonas aeruginosa* has been carried out by Modak and Fox (1973). Using silver sulfadiazine, from which silver ion is released as the major active component, they found silver to bind mainly to the cell protein and carbohydrate fractions. However, the major biological effects were caused by the interactions with the cell DNA. Inhibition was observed at 0.05 μM, and cell killing occurred at a 0.2 μM concentration. Inhibition of growth initiation was observed at a silver concentration of 0.003 μM. The DNA-bound silver concentrations increased to a peak value during growth inhibition, and growth did not begin until the Ag/DNA level had decreased to give a base pair to silver ion ratio of 80. Below a base pair to silver ion ratio of approximately 200, normal growth occurred. Although on a relative basis, less silver in the cell was associated with the DNA the DNA had the greatest relative affinity for silver in comparison with the RNA and "cell residue" fractions. Also, during both inhibition and growth, the silver in RNA and cell residues remained essentially constant. These results also emphasized that silver ion binding to DNA is both chemically and biologically reversible. Related studies using *Plasmodium bergeri*, a causative agent for malaria, have been completed by Wysor (1975a,b).

Prusik and Geacintov (1976) describe an interesting application of spectrosopic techniques to this problem. Silver and manganese ions moved from base pair to base pair along the DNA helix. The technique, which utilized optical excitation of dye in the helix, demonstrated that even small amounts of bound metal ions decrease the lifetime of the excited states of the dye. Metal ions, they noted, were present in only trace amounts in cellular DNA. In addition, this study showed that silver ion has a much lower mobility than manganese ion.

This type of study, on the high affinity of "cell residue" components for metals, has been emphasized by Beveridge and Murray (1976) who studied uptake and retention of 18 metals by cell wall preparations from *Bacillus subtilis*. Small amounts of silver (3.8 μg/mg cell walls) were taken up together with strontium, lead and mercury, whereas substantial copper was absorbed (190 μg/mg dry weight of cell wall materials). The cell wall fragments generally behaved as nonspecific ionophones, especially at the cell wall surface. In spite of earlier suggestions by Danielli (1944) and Somers (1959) regarding electronegativity and toxicity, these workers could find no simple electrostatic explanation for the binding of a metal ion to a ligand; and suggested that each metal may bind to a unique set of sites in the wall, possibly by different mechanisms. Some of the characteristics which might be considered to influence this adsorption were:

1. cation size,
2. electronegativity,
3. heat of hydration, and
4. size and/or arrangement of the hydration shell

Silver, being one of the metals taken up in small but significant amounts, had an intermediate ionic radius and heat of hydration, but these physical comparisons were not considered to be helpful in interpretation of the experimental data.

Since the pH and redox levels were not controlled in these experiments, additional interpretations are difficult; but further work in this area might be warranted.

6.6 ACQUISITION OF RESISTANCE

Another unique characteristic of microorganisms is that they possess a series of mechanisms by which they can become resistant, accommodated, or tolerant to metals which initially were toxic. This suggests that after the first addition of a metal they may respond as sensitive cells, but with a second application the system might respond differently. Recent work shows that the acquisition of metal resistance may be genetically controlled, and that exposure to a single metal may confer resistance, not only to the original metal, but to other metals.

An excellent conceptual summary has been provided by Ernst (1976), in which accommodation and resistance to metals by microorganisms are discussed. Generally, the microbial cell wall is considered to have a high affinity for some metals, and microbes can develop special enzymes which are resistant, by various mechanisms, to inactivation by heavy metals. In available work, the evolution of heavy metal-resistant enzymes is felt to be restricted to microorganisms.

Tolerance is a slightly different concept, as discussed by Ernst (1976), which can be due to the maintenance of a higher rate of sulfhydryl group synthesis at critical regions in the cell. Specifically, the formation of increased levels of nonprotein sulfhydryl groups has been known to occur in *Aspergillus niger* (Ashworth and Amin, 1964) and in yeasts (Ashida and Nakamura, 1959). Based on work with plants, the suggestion was made that "the maintenance of metal tolerance requires much energy," which might be expected, considering the large energy expenditure required to generate sulfhydryl-level oxidation state electrons.

Many general reviews are available concerning the phenomena of resistance and tolerance to metals among microorganisms. As noted by Zimmermann (1952), the first observation of silver-resistant microorganisms was made in the 1920-1930 period by Hosenfeld (1938a,b) and other workers. Repeated growth and exposure to higher silver concentrations allowed growth to occur in the presence of relatively high concentrations of metals (in the range of 50 mg/liter). Frequently the resistant strains produced colonies of a dark metallic silver color on silver ion containing media, indicating their ability to reduce ionic to metallic silver, which was generally not possessed by the sensitive original culture.

Addicks (1940) discussed the significance of microbes developing "resistance" versus being able to "adapt" or to be "trained" to tolerate high levels of particular metals, in an interesting summary of earlier work in this area. This is an example of the "premolecular biology" or "premicrobial genetics" approach to problems of interpreting phenotypic change phenomena.

More recently, the genetic basis for the acquisition of a metal reducing capability has been studied by Summers and Sugarman (1974). A mercury-reducing plasmid, located in the cytoplasm of *E. coli*, allowed this organism to bind 4 to 6 times more silver to the cell than mercury-sensitive cells. Both silver and gold ions were substrates (which were reduced to the elemental metals), as well as inhibitors of the mercury-reducing enzymes. However, incubation of sensitive cells with ionic silver or gold did not cause induction of mercury-reducing activity. Again, with the expenditure of energy, microorganisms can detoxify an ionic metal by production of elemental metals which accumulate in the cell wall membrane zone, and which will have essentially no effect on the cell in that form.

Similar acquisition of resistance or tolerance has been observed for *Hyphomicrobium* (Giangiordano, 1976) and by microorganisms in the soil environment (Wildung et al., 1976).

6.7 EFFECTS ON METABOLIC PRODUCTS AND CELL COMPOSITION

In addition to developing "tolerance" or acquiring genetic-level resistance, a microorganism can alter the types and amounts of cell products, due to the presence of silver in the environment. Khovrychev et al. (1976) studied the amino acid composition of *Candida utilis* in chemostat culture in the presence of a low silver ion concentration (2.3 mg/liter at $D = 0.1\ hr^{-1}$ and 0.15 mg/liter at $D = 0.2\ hr^{-1}$). Generally, the total amount of amino acids increased in the presence of silver, with increases in glycine, alanine, leucine, glutamic acid, and lysine observed.

Marsh et al. (1975) studied the effects of trace metals (including silver) on the production of aflatoxin by *Aspergillus parasiticus*, and noted that at a silver ion concentration of 0.2 μg/ml alfatoxin synthesis was reduced, but no effect on total biomass production occurred.

6.8 STUDIES WITH SILVER IODIDE FROM SEEDING AGENT GENERATORS

Silver iodide has exhibited antimicrobial effects, but to detect these it is necessary to use extremely high concentrations, or to use agar culture systems where dissociated silver ion can interact with susceptible microbial populations. Information provided in the Merck Index (1968) indicates that silver iodide, used as a 40% slurry, is useful as a local antiseptic on mucous membranes. Studies of the growth or the survival of selected bacterial species in the presence of seeding generator-derived silver iodide, have not shown definitive negative effects in liquid culture systems. Klein and Sokol (1976) found that the growth rate of *Arthrobacter*, growing diauxically on a mixed carbon source medium, was not decreased in the presence of relatively high (3.6 μg/ml) concentrations of seeding agent generator-derived silver iodide (Fig. 6-5). However, with higher levels of silver iodide, slight decreases in the formation of adaptive enzymes were noted (Fig. 6-6).

In a more recent study, Steven (1977) found that silver iodide from sodium or ammonium iodide seeding generator complexes had no observable effects on the viability or energy charge status of a water-derived *Aeromonas* culture over a three week starvation period. In this experiment the silver iodide was used at 10 and 100 mg/liter of starvation medium.

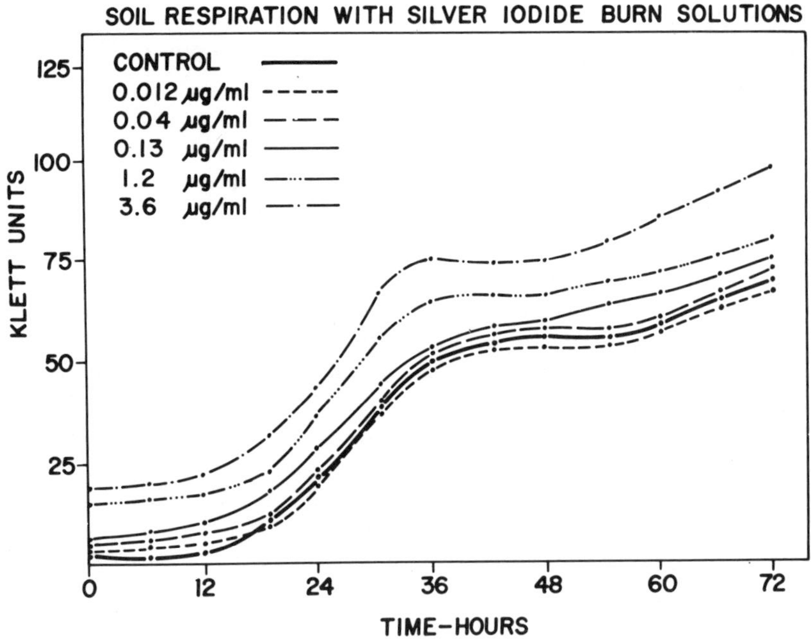

FIGURE 6-5. Arthrobacter *growth responses in the presence of varied levels of silver iodide derived from seeding generator burn solutions (Klein and Sokol, 1976).*

Castignetti (1977), in related studies, found that silver iodide from the same burn complex forms had no observable effects on methane production by a sewage-derived culture in a rigorously defined environment. In another related study, Klein and Giangiordano (1976) and Giangiordano (1976) found no effect of silver iodide from burn complex materials on *Hyphomicrobium* growth initiation over a range of temperatures, or on morphology of this prosthecate bacterium common to water, mud, and soil environments.

Some evidence for silver iodide effects on microbial activities in laboratory test systems is available. Burris (1977) found that *Chromobacterium*, a violet-pigmented organism derived from soils, showed decreased viability in the presence of silver iodide derived from reagent-grade chemicals or generator burn complexes. In a related study, the growth and spread of *Arthrobacter* through a sand matrix containing silver was monitored using the Stotzky replica plating technique (Stotzky, 1965) by Klein and Sokol (1976), where decreased growth was observed. The matrix consisted of washed sand (average particle size about 1mm) which was adjusted to 70% of moisture-holding capacity with a mineral salts medium containing 2.5 mM succinate. Silver nitrate was added to the mineral salts, prior to mixing with the sand, at a concentration of 2 mg/kg sand. A fine powder of silver iodide was added to the moistened sand and then thoroughly mixed. The silver-sand mixtures were then placed in petri dishes (six replicates per treatment), tamped down to produce a smooth level surface and autoclaved for 1 hour. After cooling, approximately 10^5 organisms were inoculated into the center of the sand plate. Periodically, replicates were inoculated into plates containing peptone-yeast extract, using a multi-needle transfer unit which could be sterilized between uses by dipping in alcohol and flaming. Decreased microorganism growth rates

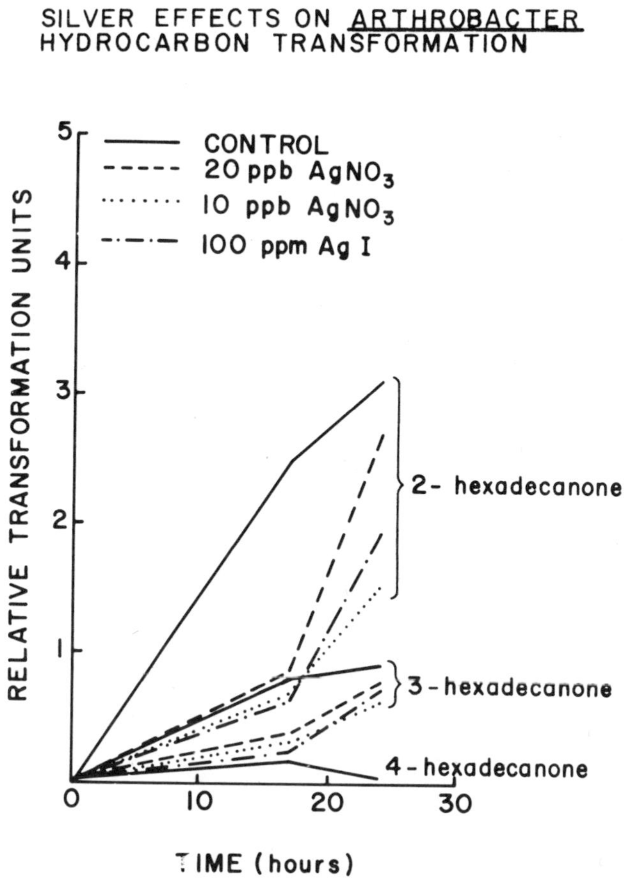

FIGURE 6-6. *Silver form and level effects on n-hexadecane transformation to isometric ketones by an* Arthrobacter *species (Klein and Sokol, 1976).*

through the sand were observed when 10 μg/g silver iodide was present (Table 6-2). Due to mixing problems, it was not possible to carry out this experiment using lower and therefore more field-relevant conditions.

Silver added to soil in the form of silver iodide derived from burn complexes is considered to accumulate and reside in the soil predominantly in that form. To test for possible interactions of silver iodide with soil microorganisms, a series of experiments using agarized soil plates and silver in the forms of silver nitrate, silver oxide, and silver iodide were completed. Nutrient agar-soil mixtures were used (surface and subsurface) at a series of concentrations, and disks of $AgNO_3$ (10 mm disks saturated with a 0.1% w/w $AgNO_3$ solution), or 30 mg amounts of silver oxide over 4 mm diameter zones were used. After incubation in the dark for up to one month at 27°C, soil microbes had caused reduction of silver ions to metallic silver. Klein and Sokol observed that silver reduction by microbial processes was stimulated in the presence of higher levels of soil organic matter (Table 6-3). This experiment also allowed observation of silver iodide inhibition of soil microorganism growth, indicating that in spite of low solubility sufficient free ion can be released, even in the presence of nonheated soil organic matter, to cause inhibition of microbial growth processes with high concentrations of silver iodide present.

This experiment was repeated using silver oxide and metallic silver, and microbial growth inhibition and silver reduction were observed with both test systems.

6.9 CONCLUSIONS

Bacterial and fungal spores appear to be more resistant to the presence of silver than vegetative forms. In addition, silver ion apparently increases the radiation resistance of bacterial spores, independent of the

TABLE 6-2. *Silver iodide effects on* Arthrobacter *growth through a nutrient broth-coated sand matrix. Growth rate in cm travelled (Klein and Sokol, 1976).*

Treatment Level		Time After Inoculation		
		24 Hours	72 Hours	144 Hours
Control	Mean	1.86	4.60	8.01
	SD[1]	0.21	0.06	0.03
AgI-10 μg/g	Mean	1.18	3.57	6.90
	SD	0.08	0.07	0.09
AgI-100 μg/g	Mean	1.31	3.45	6.82
	SD	0.11	0.15	0.16

[1]Standard Deviation

TABLE 6-3. *Nutrient agarized soil effects on silver reduction and growth inhibition effects (Klein and Sokol, 1976).*

w/w Percentages		Silver Form			
		$AgNO_3$		AgI	
Surface Soil	Sub-surface Soil	Chemical Reaction	Growth Inhibition	Chemical Reaction	Growth Inhibition
10		Silver Reduction	Pronounced	Silver Reduction	Slight
1		Silver Reduction		— —	Slight
0.1		[1]	Good 2nd Growth Band	— —	Slight
	10	vs/silver red		— —	Slight
	1	— —		— —	Slight
	0.1	— —		— —	Slight
	0.01	— —		— —	Slight
Control		— —	None	— —	None
Control		— —	None	— —	None

[1]No reaction observed

oxygen concentration. The vegetative (actively growing) forms of many microorganisms are often stimulated by low levels of silver, but are generally inhibited by the metal at higher levels. However, these effects are dependent upon the silver compound used, the particular microorganism and experimental conditions. Oxygen presence or absence can influence the antimicrobial characteristics of various forms of silver.

Silver appears to affect microorganisms at many sites, but the most common sensitive function is the sulfhydryl group. Organisms which can synthesize excess free sulfhydryl groups may develop a tolerance to silver, while those which do not will experience enzymatic inhibition. Another major site of biochemical interaction is the cell DNA, to which the silver binds reversibly at several reactive groups, such as bases, riboses, and phosphates.

Silver iodide from seeding generator burn complexes may produce antimicrobial effects, but they are not observed consistently, and only with concentrations which far exceed levels expected to accumulate in natural environments.

7

Soil Microbiological Processes

D. A. Klein

7.1 DEPOSITION STUDIES

To predict the fate and possible effects of silver derived from nucleating agents in the soil environment, several points must be considered: (a) where and how much nucleating agent is being added to soils over time (Chapter 2), (b) can one predict the forms which will eventually accumulate, based on chemical information (Chapter 3) or biological processes, (c) what factors will influence silver movement, and (d) can silver already present in soil be differentiated from that which is being added through weather modification activities?

On a global basis, silver is present in the earth's crust at only a concentration of 0.07 mg/kg. Mason (1966) noted that silver is present at a level of 0.04 mg/kg in qranites, and in diabase materials (partly crystalline rock of ophitic texture) this element will be present at 0.06 mg/kg, near the crustal average.

Boyle (1968) has provided information on the variability of silver concentrations in various crustal materials and their related soils. Values of between 0.32 to 0.37 mg/kg are given for normal schist-quartzite- and greenstone-derived soils. Over lead-zinc-silver lodes, silver levels of up to 10 to 20 mg/kg are observed. Above glacial hills and silver veins, levels of 200 mg/kg have been observed. As a component of the biosphere, Webb and Fearon (1937) as noted in Boyle (1966), consider silver to be a "variable microconstituent."

Silver tends to concentrate in surface soil zones through abiotic and biotic processes (Presant, 1963; Presant and Tupper, 1965), as noted in Table 7-1 which summarizes data from the high rainfall New Brunswick area. These workers suggested that silver will tend to combine with organic material, and silver levels in surface soils were generally five times higher than those of the parent material. A direct relationship between increased soil pH and silver retention in the surface soil horizons was noted at pH values above four. Silver levels were higher in bog soils, marshes, and other areas where organic debris had accumulated, apparently due to the strong adsorption of silver by humic colloids.

TABLE 7-1. *Silver content of soils, Bathurst-Newcastle area, New Brunswick (Presant, 1963).*

Horizon[1]	Ag (μg/g) range	Ag (μg/g) Mean of all horizons (26 profiles)	Ag (μg/g) Mean of normal horizons (14 profiles)	Ag (μg/g) Mean of horizons over Brunswick No. 6 Deposit[2] (2 profiles)	Ag (μg/g) Mean of horizons over Nigadoo Deposit[3] (5 profiles)	Ag (μg/g) Mean of horizons over Middle River Deposit[4] (2 profiles)
A_0	0.2-90.0	5.7	1.6	45.6	5.2	3.4
A_2	0.1-36.0	2.4	0.4	18.6	0.4	1.8
B_1	0.2- 6.7	1.8	0.3	3.9	4.6	4.1
B_2	0.1-24.0	3.5	0.3	19.5	4.1	6.1
C	0.1-16.0	2.7	0.3	5.3	9.0	Not detected

[1]Soils are podzols developed mainly on glacial till.
[2]Brunswick No. 6 deposit is a massive sulphide body containing essentially Pb, Zn, Cu, Fe, Ag, S.
[3]Nigadoo deposit is a sulphide vein deposit containing essentially Pb, Zn, Cu, Fe, As, Ag, S.
[4]Middle River deposit is a massive sulphide body containing essentially Pb, Zn, Cu, Fe, Ag, S.

Boyle (1968) also provided interesting information on silver movement and concentration in soils. Silver was considered to bind to clay minerals due to its similarity (in ionic radius) to potassium and sodium. In addition, through oxidation and decomposition processes, silver can concentrate above the background levels in the surface organic matter layer. Apparently silver iodide can also occur in a series of polymorphs, primarily in arid regions. In the U.S., these have been observed at Tonopah and Goldfield, Nevada; at Tulce Valley, New Mexico; and at Peavie Hills, Arizona.

When silver is deposited in soils via precipitation or during snow melt processes, it will be retained in surface soil layers. Cameron (1973) studied silver transport in soils, and noted that silver ion absorption was influenced by soil organic matter presence, while AgI adsorption was largely independent of organic matter levels. These studies also indicated that adsorbed silver is not readily released back into solution.

A wide range of field studies have also confirmed the observation of silver retention in soils. Teller et al. (1976) studied a seeding generator site at Steamboat Springs, Colorado, last used five years prior to their study, and noted that a major portion of the added silver was retained in the surface soil zone. Only at the immediate seeding generator site was movement of silver below the surface soil (approximately 10 to 15 cm depth) observed. In related studies of silver infiltration in soil cores obtained from the field (Corelli, 1971), a marked retention of silver in the upper soil zones of an Ascalon soil was also noted (Fig. 7-1), where burn complex-derived silver iodide was added at 0.5 and 5.0 mg/liter in the solution passed through the soil column. As would be expected, the highest concentrations occurred in the surface zones. Soil profile saturation did not occur with the addition of solutions containing silver at 5.0 mg/liter, which is at least 1,000 times higher than the maximum total silver values reported for seeded precipitation. In addition, there was no significant difference between the silver concentration of the test leachate and that of plain water which had been passed through control cores, confirming that silver is efficiently absorbed by soils.

Additional studies on the silver-holding capacities of soils have been carried out by Weaver (1971). In this work, known amounts of soils were swirled in solutions containing silver ion, and the decreases in silver ion in solution were measured over time. Fifteen soils were used, and these absorbed between 1 to 3% of their weight in silver, equivalent to 1×10^{-5} grams of silver per gram, or approximately 20 kg/hectare in a 15 cm plow depth. However, less silver adsorbed in the soil at lower silver concentrations, which led Freeman (1975) to suggest that absorption may not be as efficient with waters containing silver at lower concentrations.

Swanson et al. (1966) studied metal sorption (they avoided the problem of defining ad- versus ab-sorption by not using the prefix) by

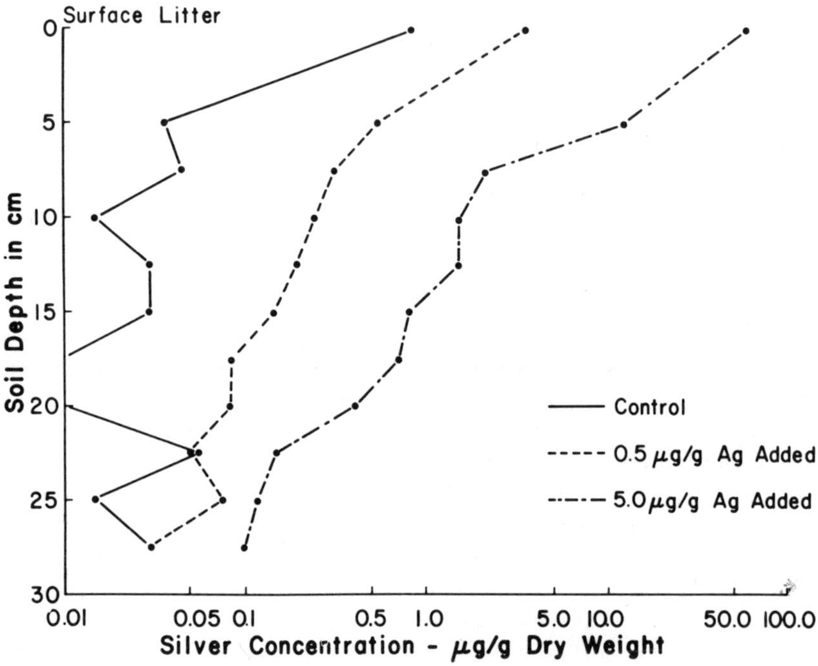

FIGURE 7-1. *The retention of added silver in surface soil (Corelli, 1971).*

northwest Florida humate materials, and both natural and chemically-extracted water-soluble organics sorbed between 1 and 17% of their own weight of a range of metals. Silver, added alone, could be sorbed to 17.4% of the humid material weight. In addition to organic components, various metal oxides (including manganese oxides) can take up silver. Anderson et al. (1973) showed that of four manganese oxide forms, the manganous manganites showed the greatest activity (up to 0.5 moles silver/mole MnO_2 at pH 7), and that this could be described with a Longmuir equation over a considerable concentration range. They suggested that the oxides of manganese and iron are the major silver scavengers during the weathering cycle.

Thus, the literature indicates that through natural processes silver from parent materials will accumulate in surface soils, and that silver derived from weather modification programs will also accumulate in this zone of intense microbiological activity.

7.2 SOIL MICROBIOLOGICAL STUDIES— LABORATORY EXPERIMENTS

Unfortunately, most work in this area has utilized silver ion, both from the point of experimental convenience and to maximize the relevance of results to the available medical literature.

Generally, when the effects of silver ion are tested, decreased microbial activity is observed quite readily in soils with levels as low as 5 μ mole/g, and silver ion has shown distinct effects on nitrogen mineralization activity (Liang and Tabatabai, 1977). As with many other studies, these workers were concerned with the possible effects of silver (and other metals) from sewage sludges. In spite of this concern they studied silver ion, which would not be the predominant silver form in such materials.

Inhibition of soil maltase by silver ion was demonstrated by Kiss and Peterfi (1960). A 0.2% solution of silver nitrate exhibited a marked effect, but complexation of the heavy metals resulted in incomplete inhibition. This study in no way indicates what would happen to soil maltase in the presence of silver iodide, since the nitrate and iodide forms tend to have markedly different properties.

In a similar study, Tabatabai (1977) found that silver ion had the greatest effects upon soil urease activity at an addition of 50 μg/g. At this treatment level, inhibition in the range of 94 to 98% was observed, depending on the soil type used.

The response of the soil microflora to the presence of heavy metals in soil was studied by Fujihara et al. (1973), and significant effects of silver ion at 1.0 μg/g on the population distribution of microorganisms, CO_2 evolution, and total aerobes, were observed; no additional details were given.

In contrast to these studies with silver ion, under short-term laboratory conditions it is difficult to show effects of silver iodide derived from reagent grade chemicals or burn complexes.

Laboratory studies using silver iodide from burn complexes (D. A. Klein and N. P. Nagle, personal communication) showed that it was not possible to detect effects upon soil dehydrogenase activity (Table 7-2). These results suggest that no effects will be observed using much higher levels of complex-derived silver iodide under short-term contact conditions.

In laboratory studies of silver-form effects on general soil respiration processes (Weaver, 1972; Klein and Sokol, 1976), distinct differences between the effects of silver ion and silver iodide on soil respiration processes also have been observed, with the silver ion showing clear effects at

TABLE 7-2. *Effects of burn complex-derived silver iodide on soil dehydrogenase activity.*

Original complex form	Nutrient addition	Absorbance at 485 nm (x ± S.D.)			
		Control	Blank filter	Filter and complex[1]	Filter and complex[2]
AgI-NaI	none	0.062 ± 0.012	0.073 ± 0.006	0.064 ± 0.009	0.079 ± 0.004
	glucose	0.007 ± 0.003	0.087 ± 0.003	0.083 ± 0.006	0.084 ± 0.006
	amino acids	0.061 ± 0	0.066 ± 0.003	0.073 ± 0.016	0.078 ± 0.019
$AgI-NH_4I$	none	0.082 ± 0.014	0.077 ± 0.014	0.083 ± 0.011	0.076 ± 0.006
	glucose	0.083 ± 0.031	0.078 ± 0.004	0.089 ± 0.003	0.082 ± 0.009
	amino acids	0.073 ± 0.019	0.078 ± 0.002	0.105 ± 0.023	0.074 ± 0.004
$AgI-TiO_2$	none	0.064 ± 0.007	0.069 ± 0.012	0.067 ± 0.006	0.058 ± 0.004
	glucose	0.070 ± 0.002	0.076 ± 0.004	0.070 ± 0.004	0.054 ± 0.004
	amino acids	0.066 ± 0.012	0.063 ± 0.011	0.062 ± 0.004	0.048 ± 0.007

[1]One 6 mm. (diameter) disc used.
[2]Three 6 mm. (diameter) disc used.
Note: Absorbance is directly proportional to μg formazan produced.

10 to 100 μg/g levels, while silver iodide has not shown such effects even at 1,000 μg/g levels under relatively short-term conditions (Fig. 7-2). As Weaver (1972) has noted, accumulation of silver iodide, even at levels much lower than tested here, would require seeding activities over extended periods of time.

7.3 SOIL MICROBIOLOGICAL STUDIES—FIELD EXPERIMENTS

Under field conditions silver from silver iodide may have more distinct long-term effects on soil microbiological processes than silver from the nitrate form, and the presence of silver in surface soils may lead to increased rather than decreased all-over microbiological activity. Sokol and Klein (1975), in studies of microbiological activity across a silver gradient imposed on an aspen-scrub community in Colorado, following operation of a ground-based generator for a period of years, found the highest microbial activities at the generator site location where the highest silver levels occurred. A positive linear correlation ($P < 0.01$) was demonstrated

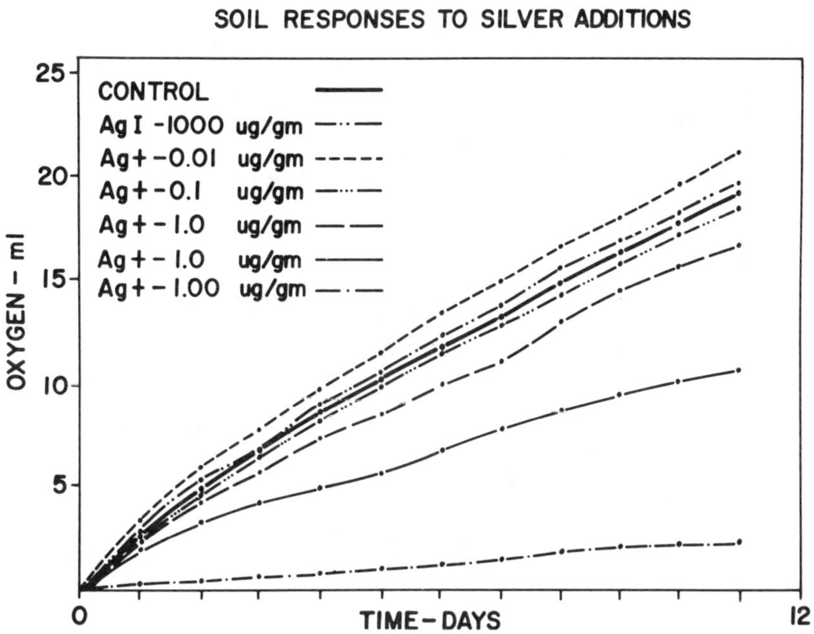

FIGURE 7-2. *Soil respiration responses to silver iodide and varied silver ion additions (Klein and Sokol, 1976).*

between organic matter content and silver concentration for the 0 to 2 cm depth soil (Fig. 7-3). The deeper soil samples show identical trends for all parameters, except that silver concentrations decreased by about an order of magnitude with each depth increase of 2 cm. Silver levels as high as 1,400 mg/kg of soil were observed in the upper 2 cm of soil adjacent to the generator site 4 years after seeding had ceased, thus confirming that added silver will tend to remain in the surface soil zone. In the areas of high silver concentrations, higher organic matter, higher water content, lower pH values, higher respiration rates, and higher microorganism populations were noted. This relationship between silver and organic matter suggests that silver presence may result in decreased organic matter turnover rates. The major weakness of this kind of study is the necessity to make assumptions on the uniformity of the plant-soil community prior to the addition of silver, which in most cases cannot be substantiated. However, this trend has also been observed at other previously-used generator sites.

Klein and Molise (1975) and Klein and Sokol (1976) carried out studies in semiarid grassland and subalpine meadow environments which also suggest, on a longer term basis, that silver iodide derived from halide complexes may have more distinct effects upon soil microbiological processes than silver originally added as silver ion (Table 7-3). Klein and Molise (1975) found that glucose mineralization appeared to be the parameter most sensitive among those tested to longer term silver presence in surface soils, and that after 3 years of contact, a direct relationship was observed between these parameters and the residual silver iodide in the surface soil layer (Fig. 7-4, 7-5, and 7-6). These results indicate that below silver levels of approximately 0.8 μg/g on an ashed soil basis, differences in glucose mineralization are no longer observable. With silver added originally in the nitrate form, no such responses were observed.

Related laboratory studies using this more sensitive technique have shown that silver iodide complexes with excess halide may have unique biological impacts which are not demonstrated by silver iodide alone. Molise and Klein (1974), using soils which had been incubated in the laboratory for up to 80 days, noted distinct decreases in glucose mineralization with silver complexed with excess iodide, and this effect was only observed after a 25-30 day incubation period (Fig. 7-7). This suggested that chemical and/or biological transformation might be required before effects of metal presence might occur.

The possibility of microbial transformation of metals has been well documented (Summers and Sugarman, 1974; Zimmermann, 1952), and may be related to the delayed silver effects observed in these studies. Generally the rate and intensity of silver reduction, a possibly important transformation step, will be greater in surface than in subsurface soils (Table 7-4). In laboratory cultures it is possible to observe reduction of

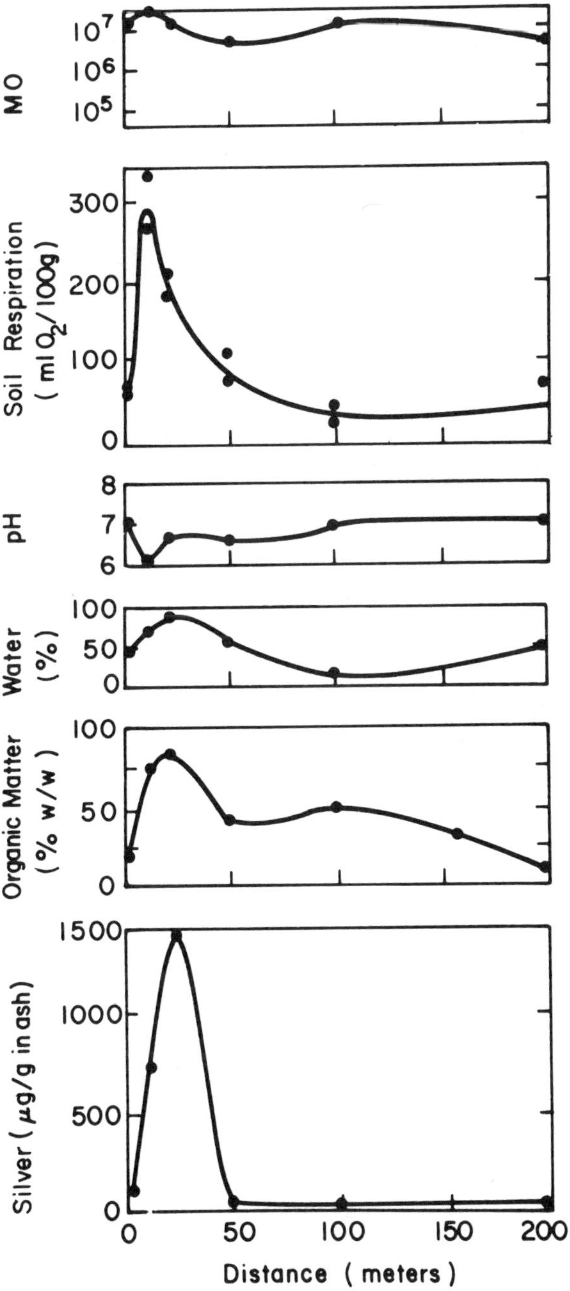

FIGURE 7-3. *Analysis of the 0 to 2 cm soil depth in a western direction from the Emerald Mountain Seeding Generator site. MO represents micro-organisms per gram of soil (Sokol and Klein, 1975).*

TABLE 7-3. *Analysis of regression between silver levels and specific biological parameters in the IBP treatment plots, 0-3 cm data (Klein and Molise, 1975).*

Parameter	Original silver form used	Relation of silver to parameter	R^2 value for curve-fitting functions				Number of samples
			Linear	Parabolic	Exponential	Power	
O_2 uptake	Iodide	Negative	0.130	0.130	0.170	0.104	41
	Nitrate	None	0.000	0.001	0.000	0.008	36
CO_2 evolution	Iodide	Negative	0.150	0.168	0.246	0.201	38
	Nitrate	None	0.000	0.000	0.003	0.001	40
Glucose mineralization	Iodide	Negative	0.215	0.295	0.317	0.494	73
	Nitrate	None	0.001	0.004	0.005	0.004	80
Organic matter-gravimetric	Iodide	None	0.013	0.013	0.001	0.006	39
	Nitrate	Positive (?)	0.197	0.198	0.166	0.060	38
Organic matter-chemical	Iodide	Positive (?)	0.000	0.136	0.001	0.034	34
	Nitrate	None	0.004	0.030	0.013	0.094	40

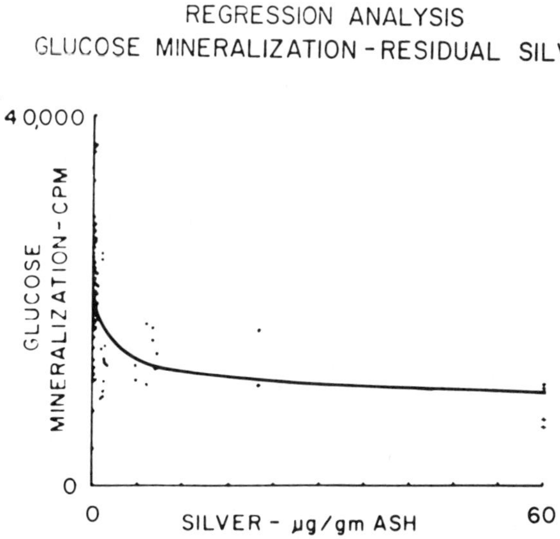

FIGURE 7-4. *Regression analysis of microbial glucose mineralization-residual silver relationships in silver-iodide treated subplots of a semiarid grassland soil. Zero to 60 µg/g samples. Linear, parabolic, exponential, and power functions R^2 values were 0.215, 0. 295, 0.317, and 0.494, respectively. The power function curve is shown (Klein and Molise, 1975).*

ionic silver to metallic silver by a wide range of microorganisms (Fig. 7-8), and also of silver iodide to metallic silver, implying that silver ion was being released during silver iodide dissociation.

With the possibility that silver iodide imposition on a semiarid grassland ecosystem is related to a slight diminution of soil respiration, a model of possible ecological ramifications of silver iodide accumulation has been suggested (Klein and Molise, 1975). Several assumptions are made in this model: (1) primary productivity will not be influenced by silver imposition, (2) silver accumulation above a certain level will result in decreased biological activities such as carbon dioxide evolution, oxygen utilization, and related plant material decomposition, and (3) the organic matter retained in the decomposer compartment will increase as a result of this decreased mineralization activity.

With these assumptions, silver added to the decomposer compartment would cause an initial decrease in microbial activity (Table 7-5). However, with the gradual accretion of organic matter through undiminished primary production, eventually the apparent microbial activity will return to the control soil value. At this "equivalence point," silver imposition effects may only be recognized by increased organic matter levels. Beyond

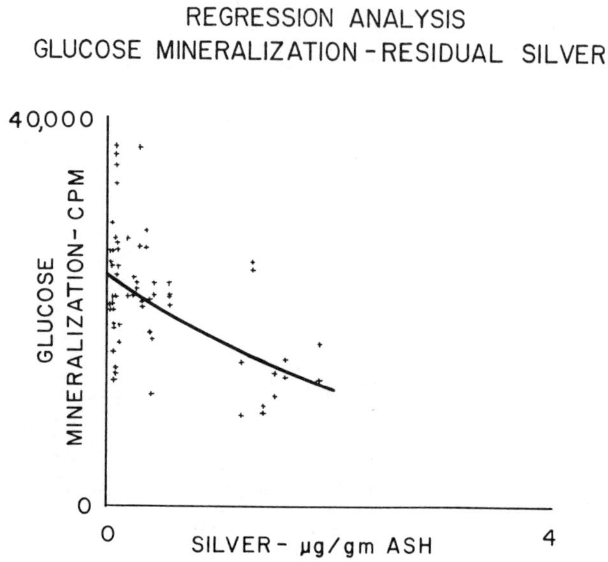

FIGURE 7-5. *Regression analysis of 0 to 2 μg/g residual silver samples. Linear, parabolic, exponential, and power function R^2 values were 0.269, 0.209, 0.325, and 0.226, respectively. The parabolic function curve is shown (Klein and Molise, 1975).*

this time, both the organic matter and microbial activity values should be higher than those observed in untreated soils.

This model predicts that possible AgI accumulation effects on decomposers may not be detectable, depending on the assay techniques used and the time of sampling. When such a biological "equivalence point" for control and treated soils occurs would depend on the particular physical conditions at the site, the rate of organic matter addition and decomposition, and the forms and level of silver present. It should be emphasized that the silver levels where these effects were first observed were at least 1 to 2 orders of magnitude higher than the background silver levels in the soils which have been examined. No significant changes in silver levels in soils or vegetation have been observed in two experimental weather modification programs carried out in Colorado (Klein et al., 1975; Teller et al., 1976), indicating that it will be difficult to establish cause-effect relationships for seeding agent accumulation under field conditions.

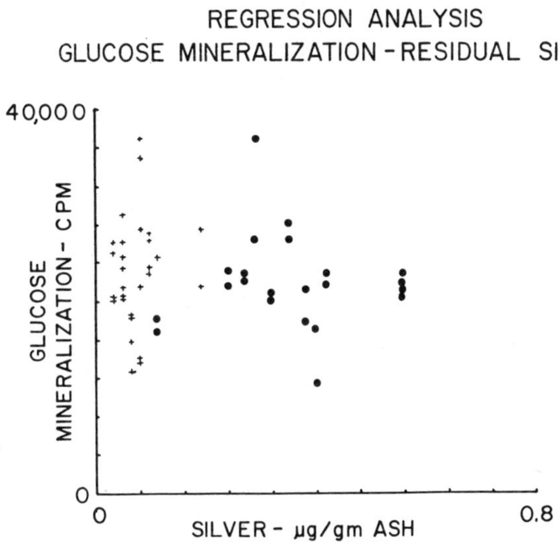

FIGURE 7-6. *Regression analysis of 0 to 0.56 µg/g residual silver samples. Linear, exponential, parabolic, and power function R^2 values were 0.007, 0.017, 0.004, and 0.002, respectively. The large circles indicate samples to which silver iodide has been added (Klein and Molise, 1975).*

7.4 CONCLUSIONS

Silver, added to snow or rainfall in the process of weather modification, will accumulate in surface soils. Subsequent rainfall will not cause major movement of this added silver below the surface soil zone. Increases in silver which may be detectable in added precipitation will be extremely difficult to detect in surface soils with available methodology, which cannot differentiate between silver forms at these low levels, and due to the relatively higher background silver concentrations in surface soils and litter materials. In addition, abiotic and biotic cyling of this metal will make it difficult to establish precise silver levels in terrestrial environments.

Soil microorganisms can potentially cause silver transformations, judging from laboratory experiments. Available laboratory and field studies show that silver added to soil in the iodide form may cause slight

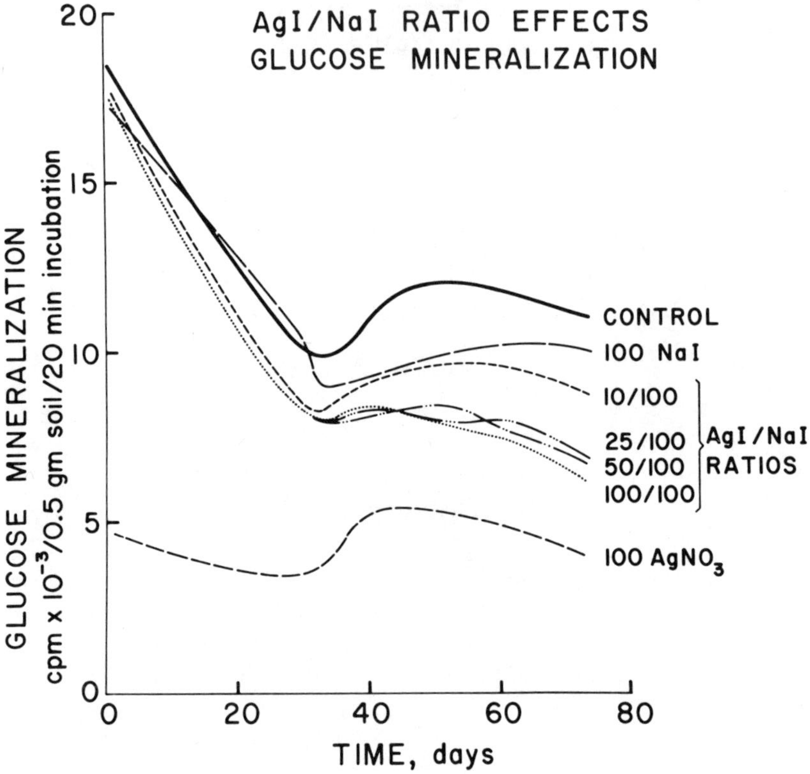

FIGURE 7-7. *Effects of silver iodide, silver iodide-sodium iodide mixtures, and silver nitrate on glucose mineralization in soils incubated under laboratory conditions (Molise and Klein, 1975).*

decreases in microbial activity under shorter-term conditions (1 to 2 months), and excess halide may have a role in this process. In contrast, silver ion can cause immediate effects on soil microbiological processes which may not be evident after longer-term (perhaps 1 to 2 year) periods. Assuming that silver presence does not decrease primary productivity, it may lead to decreased litter decomposition rates in surface soils, which may lead to the development of soils with higher rather than lower microbiological activities.

The silver levels where these effects might first be observed are at least 100-fold higher than levels which normally occur in the surface zone of the soils examined in these studies.

TABLE 7-4. *Soil effects on silver reduction and microbial growth in agar cultures (Klein and Sokol, 1976).*

w/w Percentages		Silver Form			
		$AgNO_3$		AgI	
Surface Soil	Sub-surface Soil	Silver reduction observed	Growth Inhibition	Silver reduction observed	Growth Inhibition
10		yes	strong	yes	slight
1		yes	strong	— —	slight
0.1		[1]	strong	— —	slight
	10	very slight	strong	— —	slight
	1	— —	strong	— —	slight
	0.1	— —	strong	— —	slight
	0.01	— —	strong	— —	slight
Control		— —	none	— —	none
Control		— —	none	— —	none

[1]No reaction observed.

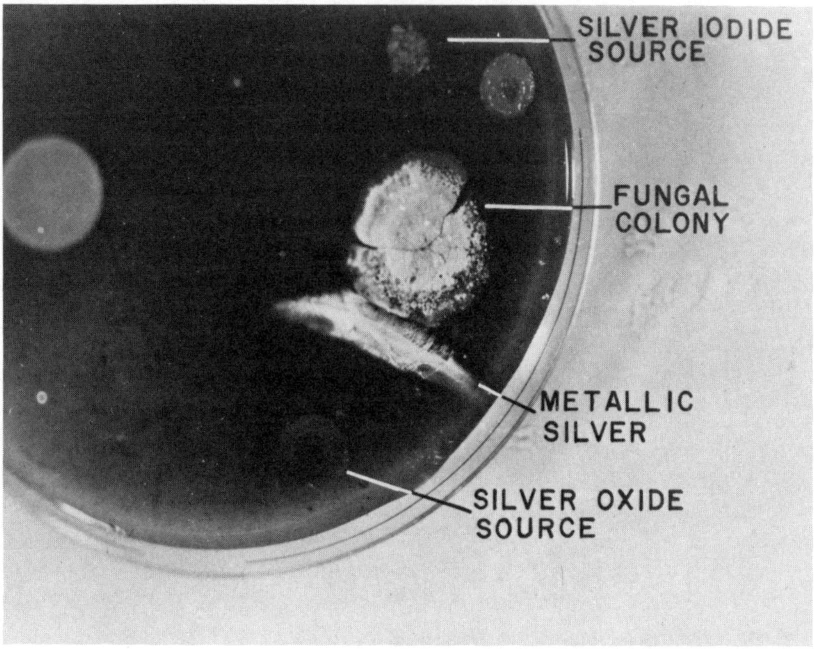

FIGURE 7-8. *Silver reduction by a soil fungal species (Klein and Sokol, 1976).*

TABLE 7-5. *Modeling of silver iodide ecological effects on decomposer functions in a terrestrial ecosystem (Klein and Molise, 1975).*

Primary productivity Annual cycle	Organic matter index with primary production input: Start	Input	Output[a]	End	Microbial activity index[b]
-1	100	10	10	100	100.00
0 (silver imposition)	100	10	9	101	95.00
+1	101	10	9	102	96.90
+2	102	10	9	103	97.85
+3	103	10	9	104	98.80
+4	104	10	9	105	99.75
+5	105	10	9	106	100.70
+6	106	10	0	107	101.65

[a] As a result of silver imposition, organic matter decomposition.

[b] Silver imposition assumed to decrease microbial activity to 95% of original, based on 100 organic matter units. This percentage activity per unit organic matter present assumed to remain constant over time.

8

Plant Effects

*D. A. Klein**

8.1 INTRODUCTION

In evaluating possible silver impacts on plants it is important to be able to determine the levels of silver in various plant parts, so that comparisons with levels in soils can be made. Regarding this concern, Boyle (1968), in a review on the geochemistry of silver, stated that "the silver content of plants varies over a wide range and hence average values are practically meaningless." As it is necessary to have some type of information on metal levels in plants, careful analyses and statistical evaluations must be used to provide an index of sampling variability. To be most meaningful, the data should be expressed on a dry weight basis, and preferably on an ash weight basis, with percent ash data being available.

Plants which accumulate silver or other precious metals have also been used in prospecting and as toxicity indicators in relation to urban and mining pollution. Antonovics et al. (1971), in an article on heavy metal tolerance in plants, discussed toxic levels of metals in soil and listed various plants that could be used as heavy metal indicators, including *Erigonom ovalifolium*, a member of the family Polygonaceae, used as a silver indicator for prospecting in Montana. *Alsine verna* (Europe) and *Lonicera cenfusa* (Australia) were also cited as indicators of silver-containing soils.

As plants can develop silver tolerance under natural conditions, most work on short-term contact effects has used greenhouse materials, or silver treatments have been imposed on existing plant communities.

*Major contributions to the completion of this chapter were made by Gayle Robison.

8.2 LABORATORY STUDIES

Many studies of silver movement and deposition have been carried out under relatively controlled greenhouse conditions. Wallace (1971) grew soybean seedlings in soil containing ^{110}Ag. One variety of soybean plant accumulated twice as much ^{110}Ag as another, which paralleled their respective ability to accumulate iron, manganese, and zinc (Table 8-1).

Pettersson (1976) found that silver was not translocated in measurable amounts to shoots of cucumber, wheat, and tomato plants which were grown for up to 15 days in a nutrient solution containing 10 μM $AgNO_3$ (Fig. 8-1). Silver retarded root growth in cucumber shoots, and the author noted that silver ion was reduced to metallic silver in the root zone, where a black precipitate was observed. Heavy metals in general accumulated to a greater extent in root than in shoot materials.

Koontz (1977, unpublished) grew bean plants hydroponically under controlled conditions, and 0.05 μM $AgNO_3$ in the nutrient solution caused a substantial decrease in the growth rate within a few days. In comparison, a nutrient solution saturated with AgI had no discernable effect. When the plant was grown with 4 liters of a 0.05 μM $AgNO_3$ solution, the silver concentration was reduced by 50% within 1 hour and no significant effect on growth was noted. With 20 liters of solution, a 50% reduction in the silver concentration took more than 12 hours and resulted in a substantially reduced plant growth rate. Adsorption by the glass container was nil. Most of the Ag was associated with the root, confirming the results of Pettersson (1976). The Ag translocated to the shoot was uniformly distributed, as shown by autoradiography. Also, Ag was exported from the leaf. Corn, on the other hand, accumulates Ag in the leaf margins (Sayre, 1952). Suzuki (1958) found only small amounts of Ag taken up by the mint plant, indicating that silver uptake rates may vary with the plant species being examined.

Weaver and Klarich (1973) determined silver uptake and plant growth of wheat, maize, and soybeans grown in sandy and loamy soils

TABLE 8-1. *Stem and leaf content of ^{110}Ag for soybean plants grown for 12 days in contaminated soil (cpm/g dry weight) (Wallace, 1971).*

DTPA Treatments	Hawkeye Soybeans		P154619-5-1 Soybean	
pp2m	Stems	Leaves	Stems	Leaves
0	5608	7730	2355	3279
10	4883	7900	2558	3826
100	2929	5577	2093	2828

Means of 3 replications.

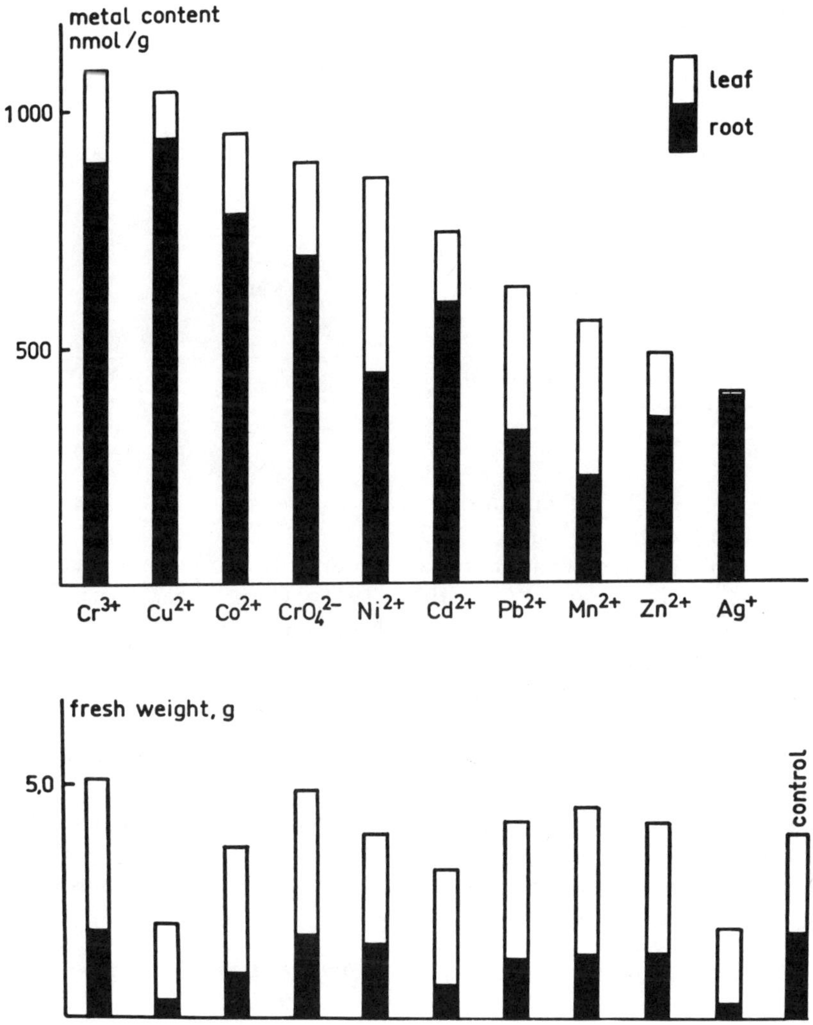

FIGURE 8-1. *Heavy metal concentration and fresh weights of cucumber plants grown for 12 days in nutrient solutions with a metal ion concentration of 10μM (means of five replicates with five plants each) (Patterson, 1976).*

enriched with silver iodide and silver nitrate. Little silver was detected in shoots of plants grown in soil enriched with silver iodide, but small amounts of silver were transported to maize shoots grown in soil enriched with soluble silver nitrate (Table 8-2). Weaver and Klarich tentatively con-

TABLE 8-2. *Silver contents (μg/g) of plants grown in soils enriched with silver[1] (Weaver and Klarich, 1973).*

	Loam		Sand			
Silver μg/g	0	1000	0	1000	Reps	Age[2]
I. Silver Iodide:						
Maize	1.03	0.82	0.75	0.80	2	10
	1.94	2.02	1.79	0.73	5	7
Soybean	0.69	0.87	0.88	0.61	2	10
Wheat	0.53	0.58	1.07	5.91	2	10
II. Silver Nitrate:						
Maize	0.72	1.02	0.41	0.44	3	7

[1]No treatment differences were statistically significant (99%, 95%).
[2]Age of plants in weeks.

cluded that even larger amounts of AgI in soils have little or no effect on crop plant growth.

Greenhouse studies have been carried out to measure silver uptake by crop plants (Teller and Klein, 1973, 1974). Corn, sugar beets, and Kentucky bluegrass were grown in washed river sand containing either $AgNO_3$ or AgI, harvested after either three or four months, and separated into roots, stems, leaves, and cobs (in the case of corn) as summarized in Table 8-3 and Fig. 8-2. Silver tended to concentrate either in or on the roots, with sugar beet roots containing from 54 to 99 μg/g silver (on a dry weight basis) when growth in the AgI treated soil. Kentucky bluegrass had slightly lower silver concentrations when grown in AgI-containing soil, but with $AgNO_3$ treatments, concentrations were 10 to 20 times higher. Silver from silver nitrate tended to move to aboveground plant portions to a

TABLE 8-3. *Summary table of silver concentration in experimental crops (μg/g in ash)[a] after harvesting (Teller and Klein, 1973).*

		Treatment			
		$AgNO_3$	AgI Burn	Control	Mean
Sugar Beet	Roots	833	70	3.3	302
	Stems & Leaves	177	8.6	1.5	62
Kentucky Bluegrass	Roots	1,015	43	1.2	353
	Stems & Leaves	170	7.1	1.3	59
Corn	Roots	728	176	3.4	302
	Stems & Leaves	26	6.5	1.2	12
	Cobs	1.66	1.03	1.2	1.3

[a]The mean values of silver in μg/g dry weight for the $AgNO_3$, AgI and control soils were 26, 12 and 0.09, respectively.

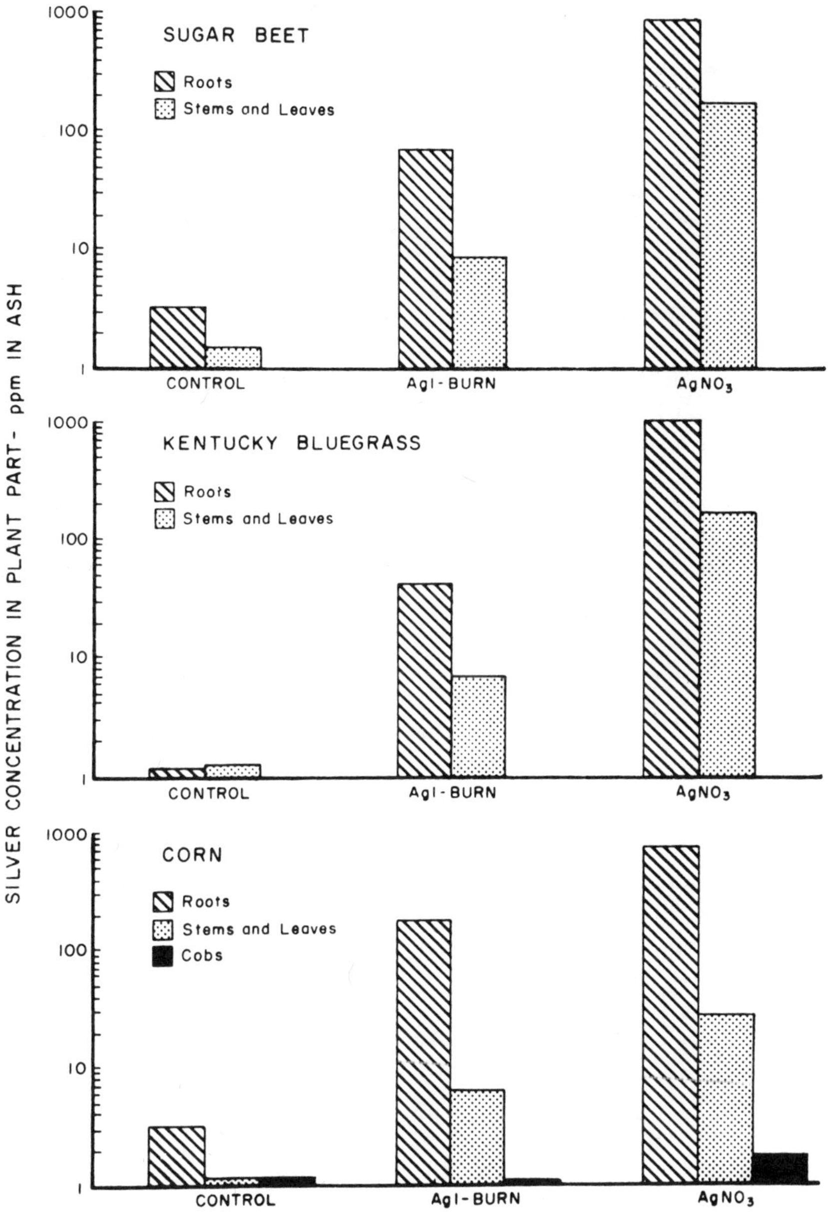

FIGURE 8-2. *Concentration of silver in corn, sugar beet, and Kentucky bluegrass from AgI burn complex and $AgNO_3$ (Teller and Klein, 1974).*

greater extent that from silver iodide, as would be expected. In the same study (Klein and Striffler, 1976), the possible location of silver in or on the roots of plants grown in silver-containing soils was examined (Table 8-4). Water washing of roots removed little silver, and washing with nitric acid (O.1 N concentration) for 30 minutes removed less than 15% of the root silver, indicating that the silver is firmly bound to, or incorporated in the root tissue.

8.3 FIELD STUDIES: GRASS AND HERBACEOUS SYSTEMS

Horowitz et al. (1974) analyzed the trace element levels of 35 plant species from 16 divisions, using field-derived materials. Silver concentrations in the plant materials ranged from 0.01 to 16 μg/g where the corresponding soils had an average silver level of 0.1 μg/g. The author noted higher silver contents in some lower plant species; in addition, 5 to 50 times more silver was found in the plants in May than in September, indicating that seasonal silver movement may occur in plants. This phenomenon will make it extremely difficult to monitor the relationship between silver accumulation in aboveground plant parts and weather modification programs.

As noted by Teller et al. (1976) for the San Juan Ecology Project, and by Klein et al. (1975) regarding the National Hail Research Experiment, no changes in the silver levels of soils, waters, or vegetation have been noted which could be specifically linked to seeding activities.

Under field conditions silver can accumulate predominantly in the root zone. In natural grass samples from northeastern Colorado, Klein et al. (1975) found that at lower silver levels, the silver was more evenly distributed between the roots and aboveground vegetation (Table 8-5). Only with higher silver levels in the soil will a relative exclusion of silver from the

TABLE 8-4. *Effects of plant root washing on residual root silver contents (Klein and Striffler, 1976).*

Washing Technique	Control Soils		Silver-Treated Soils: 70 μg/g		Silver-Treated Soils: 144 μg/g	
	Ash %	μg/g	Ash %	Ag μg/g	Ash %	Ag μg/g
Root and Soil Control	75.0	4.3	—	—	78.6	255
Water Washing	59.7	6.2	68.5	130	41.6	640
HNO_3-10 min.	35.2	2.4	53.0	290	26.3	780
HNO_3 20 min.	37.7	2.8	26.4	235	26.4	1050
HNO_3-30 min.	44.8	1.5	32.0	250	24.8	860

TABLE 8-5. *Silver distribution in plant-soil ecosystem components as a function of silver level in the system[a] (Klein, 1975).*

		Silver Content μg/kg in Ash[b]				
Sample	Number	V	CS	RS	SS	R
Silver range						
ppb <32	4	26 ± 9	13 ± 5	24 ± 16	12 ± 7	21 ± 12
	10	40 ± 10	33 ± 22	28 ± 13	25 ± 17	37 ± 13
74-95	11	63 ± 15	42 ± 19	43 ± 19	48 ± 18	44 ± 14
96-210*	5	112 ± 21	40 ± 10	50 ± 19	40 ± 20	80 ± 30
<220*	4	280 ± 49	50 ± 19	74 ± 24	58 ± 20	192 ± 41
All Samples	34	83 ± 76	37 ± 19	42 ± 21	40 ± 21	63 ± 54

[a] Asterisk (*) indicates significant differences in silver distribution observed between subcomponents of the soil-plant system, based on standard deviation ranges.
[b] = number of samples in the given silver concentrations range
V = vegetation
CS = complete soil
RS = roots plus soil remaining after sifted
R = roots

aboveground plant portion occur. Klein and Sokol (1974) observed increased silver exclusion from aboveground plant parts with higher silver levels in treated soils (Table 8-6). Thus, less silver is taken up by the plant on a relative basis, with increasing silver levels.

Giordano and Mortvedt (1976) studied the effects of nitrogen addition on uptake and mobility of heavy metals in tall fescue grown in sewage

TABLE 8-6. *Silver level and form addition effects on soil and grass silver content in a subalpine meadow community (Klein and Sokol, 1976).*

Treatments	Silver in Soil (μg/g)	Silver in Grass (μg/g)	Ratio Plant Silver / Soil Silver
Controls — untreated	0.028	0.18	5.04
$AgNO_3$ Solvent Control	0.096	0.04	0.42
AgI Solvent Control	0.057	0.064	1.12
Treatments:			
$AgNO_3$ 1 (μg/g)	0.08	0.150	1.88
10 (μg/g)	4.000	6.70	1.68
100 (μg/g)	127.100	7.80	0.06
AgI 1 (μg/g)	0.110	0.07	0.62
10 (μg/g)	6.140	0.55	0.09
100 (μg/g)	43.800	4.90	0.11

sludge-treated soil. Although silver was not measured, the Zn, Cd, Cr, Pb, Cu, and Ni levels considered might apply to the measurement of silver uptake by plants. Nitrogen fertilization exhibited no specific effect on metal uptake, but overall uptake was enhanced by increased plant growth.

Mortvedt and Giordano (1975), in another study, reported on corn response to Zn and Cr in wastes applied to soil. Their research showed that Zn in municipal wastes is available to crops; however, uptake was less than 5% of the soil concentration. Chromium in waste products was not available to crops, paralleling results found with silver.

There is a lack of consistent interest in the possible fate of silver, from sewage sludges, in the plant-soil system, suggesting there is little concern regarding its effect on bioconcentration or human toxicity.

Silver movement in the plant-soil environment has also been examined in plant communities surrounding previously used seeding generator sites (Teller and Klein, 1974). Silver is retained in the surface soil zone, as illustrated in Fig. 2-2 and Table 7-3. They observed a distinct movement of silver into litter, grass, and aspen foliage (Table 8-7). A more detailed analysis of data from an aspen community (Teller and Klein, 1973) revealed that plant tops have higher silver concentrations (Table 8-8) due apparently to translocation rather than rainsplash. In spite of the relatively higher silver levels in these aspen plants, no obvious effects upon plant vigor were observed.

8.4 USE OF TREES AS SEEDING AGENT MONITORING SYSTEMS

The potential for monitoring long-term heavy metal pollution by using trees was examined by Sheppard and Funk (1975). The authors cite earlier work by Warren, who found heavy metal levels were 10 times higher in trees growing near ore deposits. In Sheppard's study, cores were taken from Ponderosa pines growing beside the Spokane River, Idaho, and metal concentrations in the dry wood were determined. Silver levels at Corbin Park were highest in rings formed from 1930 to 1932. A low of 0.01 μg/g was found in rings formed sometime between 1971 and 1975. The mean value for silver was 0.12 μg/g. The silver level changes in the trees appeared to follow changes in mining activity, with a delay of about four years. The authors suggested that trees located close to rivers can be used to monitor heavy metal concentrations in river water.

Lepp (1975) also emphasized that trees can be useful for monitoring heavy metal pollution by three means. Root uptake was considered to be important, but dependent on pH, with a high pH limiting uptake. Foliar uptake, another important mechanism, can occur when soluble metals penetrate the leaf cuticle. Jensen and Kavaljian (1956) found that foliar-

TABLE 8-7. *Silver concentrations in litter, grass and aspen foliage at Steamboat Generator Site, 4 years after last seeding (Teller and Klein, 1974).*

Dist. from Ground Zero (meters)	Direction	Silver Concentration, μg/g dry weight		
		Litter	Grass	Aspen Foliage
0	—	17.0	54	3.6
10	N	220	5.0	1.3
	S	105	5.6	2.2
	E	205	3.3	1.6
	W	210	6.6	1.6
20	N	150	2.3	3.7
	S	31	5.2	2.5
	E	1120	12.0	0.72
	W	610	14.0	3.4
50	N	660	5.8	3.2
	S	23	3.2	0.96
	E	185	2.9	0.70
	W	180	27.0	2.8
100	N	90	1.6	1.8
	S	5	0.66	0.85
	E	28	0.80	0.35
	W	4.5	3.0	1.2
200	N	36	0.40	0.55
	S	0.88	0.064	0.26
	E	9.6	0.32	0.096
	W	1.0	1.1	0.80
500	N			
	S	0.64	0.35	0.14
	E	1.0	0.096	0.13
	W	1.8	4.8	0.80

absorbed metals were transported via the phloem, but loading barriers could have a major influence on this mode of transport. Direct stem uptake, although not well understood, appears to be important in deciduous

TABLE 8-8. *Silver concentration in young aspen foliage (Steamboat Generator Site)*—μg/g silver in ash (Teller and Klein, 1973).

Tree Component	Height Above Ground Level	
	1 Meter	2.5 Meters
Foliage	13.0	25.0
Twigs	5.3	15.0
Stem Bark	3.2	13.0
Stem Wood	2.4	22.0

trees where stem deposition would increase during the winter. Stem transport was indicated to be subject to few limiting factors, although it was apparently slow. It was noted that lateral transport may have an effect on heavy metal levels of growth rings. In addition, root-absorbed metals were considered to eventually be distributed more completely through the plant, possibly making interpretation of such data difficult.

Tiffin (1975) reported on the form and distribution of metals in plants and provided some basic concepts. He considered that plants may act as passive metal receptors through fallout or root absorption, and may also modify the form of metals, as in metal reduction in root cap cells. Internal distribution may occur by localization in stems or roots, and by binding metals to cell walls in metal-tolerant plants and plants that can tolerate accumulation, without toxic effects, as these are present predominantly in the less reactive elemental form following reduction.

Geographic location may also influence the metal content of a plant. Ellis (1962) observed a mean of 0.21 μg/g silver in trees east of the Cascade Mountains, with a range from 0.02 to 1.52 μg/g. However, in trees west of the Cascades, the mean was 0.02 μg/g, with levels ranging from a low of 0.01 μg/g to a high of 0.07 μg/g. Needles from grand fir trees east of the Cascades had very high levels, averaging 28.7 μg/g in dry tissue. Sound wood from the same area had only 0.7 μg/g. This may implicate changes in rainfall, and also more general climatological factors, as influencing silver movement in plant communities.

In spite of geographical (and climatological) effects on silver levels in tree rings, it may still be possible to monitor silver imposition within a watershed, assuming a general background accumulation level can be established. In another interesting study, Jaffre et al. (1976) studied a tree of New Caledonia (*Sebertia acuminate*) whose latex contained over 25% nickel (dry weight). This hyperaccumulator, which contained 11% nickel wet weight on a total plant basis, caused considerable interest due to its possible association with nickel deposits. In addition, it posed interesting problems in plant physiology regarding tolerance of a normally toxic element. It may be possible that there are other plant species with a similar ability to concentrate silver, which might make them suitable for monitoring purposes.

8.5 PHYSIOLOGICAL EFFECTS OF SILVER ON PLANT GROWTH PROCESSES

Much of the available general literature in this area up to 1975 has been summarized by Smith and Carson (1977). Unfortunately, silver nitrate has been the silver form used in most of these studies, making them difficult to apply to weather modification impact assessment. However, several field and laboratory studies have been completed in the last few years in which silver iodide or silver iodide derived from burn complexes were used, with the specific objective of obtaining information on potential impacts of nucleating agents on plant communities. Weaver (1972) used silver iodide at 0, 10, 100, 1,000, and 10,000 μg/g, and evaluated plant growth in pot cultures. Silver iodide at 1,000 μg/g had little effect on growth of the plants studied (wheat, maize, and soybeans). Wheat growth was initially reduced in sand containing 100 to 1,000 μg/g silver iodide, but over the total experiment neither wheat nor maize was significantly affected by up to 1,000 μg/g silver iodide in either sand or loam. Enrichment of soils with silver nitrate, however, caused a distinct reduction in maize growth.

Singleton and White (1975) studied the effects of silver iodide additions to soil on the emergence and germination of tufted hairgrass or lodgepole pine. Silver exhibited a greater inhibition of emergence than of germination for both plants studied, but even at 60 μg/g AgI, at least 15% of grass seedlings and 20% of pine seedlings emerged (Fig. 8-3). The authors concluded that the addition of AgI would probably not have any undesirable effects on plants or soil except possibly in the immediate vicinity of AgI generators.

Fins (1972) found that silver nitrate inhibited both germination and pollen tube growth in Engelmann Spruce, while silver iodide tended to reduce pollen tube length at silver levels at least 10^4 higher than might occur in seeding precipitations. Silver iodide at 1.4×10^3 μg/g inhibited germination, while at lower levels a trend toward stimulation was observed. Silver iodide-sodium iodide, silver iodide-ammonium iodide, and silver iodide from burn complexes did not affect germination, and slight stimulation of pollen tube length occurred at the lower concentrations, as noted with silver iodide alone. With silver nitrate, as might be expected, germination was inhibited at relatively low levels (1.0×10^{-7} μg/g of test medium), while a stimulation-inhibition pattern occurred as the silver ion concentration was increased. Obydennyi (1976) found that silver nitrate stimulated growth of evergreen and deciduous trees and shrubs in areas of industrial air pollution (no details were given), indicating that metal presence at low levels may lead to stimulation as well as inhibition.

Silver ion, although known to be toxic when present in soils used in plant growth studies, can have beneficial effects on tomatoes and peas when applied by foliar treatment by serving as a potent inhibitor of ethylene-induced epinasty, senescence, and abcission. In studies with peas, Beyer (1976a) found that foliar application of silver ion prevented leaf abcission when used in 50, 100, and 200 μg/g solutions one day prior to a 7-day exposure to 12 μl/liter of ethylene. In addition, no major growth differences were observed in control plants growing in an ethylene-free atmosphere with and without silver ion sprays at 200 μg/ml. Silver ion was also

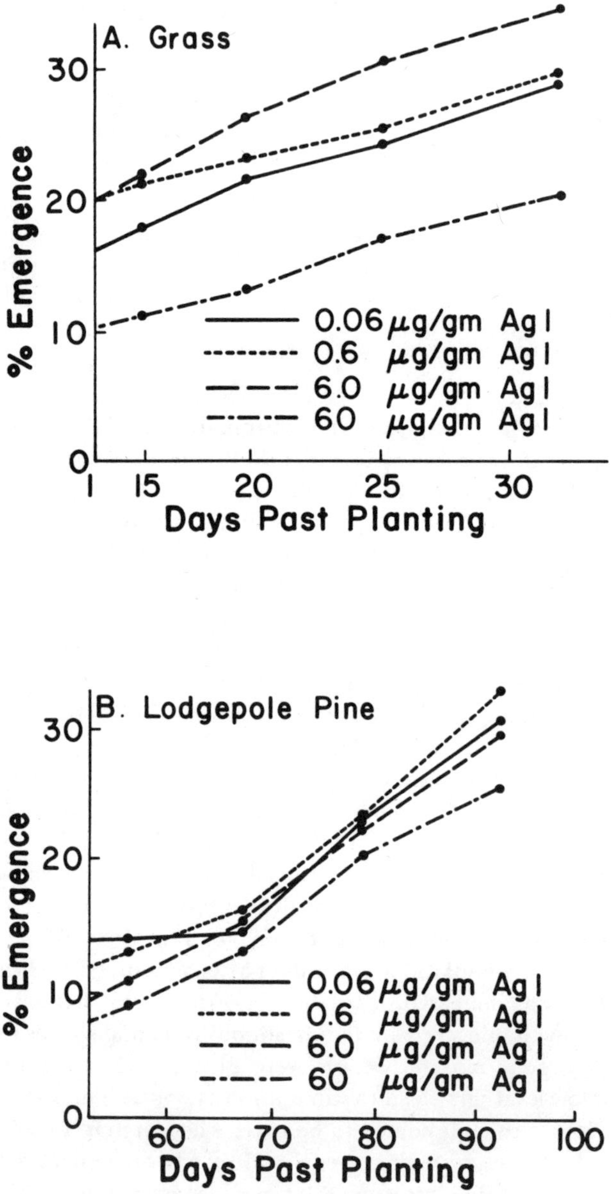

FIGURE 8-3. *The percentage emergence of grass (A) and lodgepole pine (B) at days past planting for four AgI treatments (Singleton and White, 1975).*

found effective in preventing ethylene-induced senescence of *Cattleya* orchids. With silver nitrate at 10 μg/ml, no damage occurred to flowers, although at 15 μg/ml some damage was noted. Similar results were obtained with cucumbers and tomatoes (Beyer, 1976b). Silver ion was considered to be an outstanding antiethylene agent due to its "persistence, specificity, and its lack of phytotoxicity at effective concentrations" (Beyer, 1976a). Silver ion was postulated to substitute for copper ion, which inhibited ethylene oxidation and hence ethylene action. These results suggest that silver, even in the ionic form and applied at relatively high concentration to plants by the foliar route, may not have major negative effects upon plants, and may actually counteract the negative effects of microbiologically-produced ethylene, which would be a more significant problem in water-logged soils.

8.6 CONCENTRATION OF SILVER IN GRAIN AND PLANT SEED-DERIVED FOODSTUFFS

A point of interest is the possible increased export of silver in food materials, primarily grains and truck farm crops in areas where summer weather modification programs might be carried out.

Kent (1942) studied tin and silver occurrence in wheat and its milling products. He found a range of silver concentrations from 0.02 μg/g in flour to 0.14 μg/g in beeswing (pericarp), and silver concentrations were only half those of lead or tin (Table 8-9). Since these data are on a fresh-weight basis, it is difficult to evaluate the actual concentrations, but this does provide an indication of silver burdens in this major food source. Bowen (1966) noted that corn (*Zea mays*) can concentrate radiolabeled silver in the leaf edge portions of the plant, and that in general, silver con-

TABLE 8-9. *Mean lead, tin and silver concentrations in wheat and wheat products (Kent, 1942).*

Sample	Lead μg/g	Tin μg/g	Silver μg/g
Flour	0.09 (7)	0.12 (3)	0.03 (3)
Germ	0.15 (4)	0.04 (3)	0.08 (3)
Bran	0.30 (8)	0.16 (7)	0.09 (7)
Beeswing (Pericarp)	0.48 (1)	0.39 (1)	0.14 (1)
Whole wheat:			
Estimated	0.10 (10)	0.09 (7)	0.04 (6)
Calculated	0.11	0.13	0.04

The numbers of samples averaged to give the above figures are shown in parentheses.

centrations in seeds can be higher than in other aboveground plant portions.

Based on these considerations, it may be valuable to monitor the silver levels in crops raised in areas where weather modification programs (especially summer programs) might take place. As crop samples produced over a period of several years can be analyzed, it should be possible to obtain information on metal background levels in crops grown in particular regions.

8.7 PHYSIOLOGICAL ASPECTS OF METAL TOLERANCE

Ernst (1976), in a study of the physiological aspects of plant tolerance to metals, found that tolerance to one metal may not necessarily confer tolerance to others. Plants have a series of mechanisms which can minimize metal effects. Avoidance occurs when the plant can excrete organic compounds which react with metals, making absorption slower than with the free metal ions. In a similar way, the formation of calcium and phosphorus compounds can also lead to reduced absorption, however, plants cannot completely prevent metal ion uptake by this mechanism. Tolerance, whereby metals are taken up but excluded from sensitive sites, can occur by the synthesis of higher levels of free sulfhydryl functions which can react with the metal ions, or by alteration of metabolic pathways. As evolution of heavy metal-resistant enzymes is restricted to microorganisms, this mechanism of tolerance is not available to plants. An especially important point made by Ernst (1976) was that much energy is required to maintain metal tolerance.

Another high energy-expenditure alternative which plants can use to minimize the effects of metal ions is to reduce them to their less active elemental forms. Brown et al. (1962) found that silver reduction occurs in association with chloroplasts but not within grana. They concluded that respiration-related processes in chlorophyll were not directly involved in silver reduction, but that ascorbic acid was the major reducing agent. Jensen and Kavaljian (1956) found high concentrations of ascorbic acid in the root cap and elongating cells, and reduction to metallic silver by the high levels of ascorbic acid at the cell surfaces, would minimize translocation.

Weier (1938) found that silver is reduced either as granules within or uniformly along the edge of chloroplasts; an observation first noted by Molisch (1938). Reduction did not occur in leaves exposed to toluene, formaldehyde, chloroform, or extreme temperatures, apparently due to destruction of ascorbic acid. Studies of silver nitrate reduction by lemon and pineapple leaf tissues further substantiated the conclusion that ascorbic acid is the major agent responsible for silver reduction in plants.

Another interesting aspect of potential silver impacts deals with mycorrhizal relationships, for which some information is available. Harley and Jennings (1958) studied the effect of silver on the respiratory response of beech mycorrhizae in the presence of glucose. Application of silver salts resulted in decreased phosphate consumption and increased ADP release and respiration. It was found that $AgNO_3$ inhibited hexose absorption between 10^{-6} and 10^{-4} M (Fig. 8-4). At a 10^{-4} M concentration, absorption was reduced almost 90%. The process of hexose inversion was much less sensitive to $AgNO_3$, and actually increased between 10^{-5} and 10^{-3} M. This study suggests that mycorrhizal relationships between plants and root-associated fungi may be useful as a monitoring method, where the rates of phosphorus transfer from the soil to the plant could be measured.

8.8 PLANT-MICROBE INTERACTIONS EFFECTS ON NUCLEANT PROPAGATION

The possibility of monitoring nucleating agent accumulation on plant surfaces by monitoring the changes in microbial population has also been

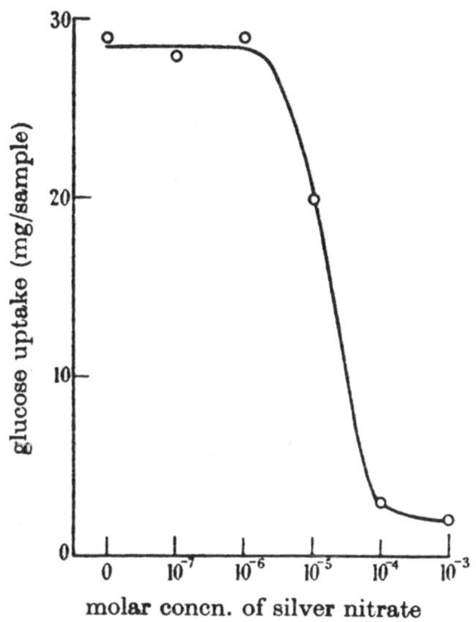

FIGURE 8-4. *The effect of silver nitrate on the uptake of glucose from 36 mM sucrose (23 hours at 20°C) by beech mycorrhizae (Harley and Jennings, 1958).*

suggested, and there appears to be information available which would substantiate the validity of this approach.

An investigation of heavy metal effects on leaf surface (phylloplane) microflora was conducted by Gingell et al. (1976). Cabbage plants and pine saplings exposed to surface contamination by zinc, lead, and cadmium from a nearby smelter were compared to uncontaminated plants. The heavy metals on the plant leaves near the smelter were present at the same ratios as heavy metals in the air. Artificial contamination was also used to simulate pollution. Levels of microorganisms on contaminated leaves were 10 to 25% of those found on controls, but the microorganisms on contaminated leaves were found to be much more tolerant to heavy metals. In addition, artificially contaminated leaves were found to be less affected by some plant pathogens than were control leaves.

Other available information (Dickinson et al., 1975; Dickinson and Preece, 1976) also emphasizes this point. In the first cited study, it was noted that in specific plants, the predominant culturable microorganisms were of the genera *Chromobacterium, Corynebacterium, Pseudomonas*, and *Xanthomonas. Chromobacterium*, an easily identifiable organism due to its purple pigmentation, has been studied for its possible utility as an indicator of transient presence of silver iodide from burn complexes (Burris, 1977).

Other studies show that plants can play a role in nucleation phenomena. Beauford et al. (1977) reported that vegetation releases particles containing metals into the atmosphere. Laboratory studies indicated that pea plants and pine seedlings released zinc and lead, but no mention is made of silver. The air surrounding the plants grown in liquid culture containing radio-labeled zinc or lead, was filtered and then condensed. The amounts of metals released and the concentrations in the plants were found to be a function of the metal concentration in the nutrient solution. The particles associated with the metals were postulated to be either epicuticular waxes lost during rapid leaf expansion, or salt crystals generated during rapid transpiration, and were less than 1 μm in size.

Field plot experiments were conducted by Beauford et al. (1977) to determine if larger particles might be involved in release of metals under natural conditions. Plots were cleared of vegetation, treated with radiolabeled zinc and lead, and left to recover their vegetation. The new foliage contained from 1 to 10 μg/g dry weight of metal, and most of the radioactivity collected in the air over plants growing in these plots was associated with particles larger than 5 μm. In addition, more zinc was released than lead. Since silver can behave similarly to zinc, silver on plant surfaces could be released into the atmosphere in a similar manner, which might result in secondary nucleation phenomena.

Atmospheric nuclei may be produced by plants in a variety of ways. Fish (1972) hypothesized that the blue haze commonly observed above forests may be due to wax particle release from pine needle tips. Electrical charges could induce production of wax particles of less than 0.6 μm in diameter. The ramifications of this phenomenon, while not immediately evident, are important. This worker suggested that silver iodide from seeding may be locally retained by plants and released to the atmosphere at a later time. The "memory effect" observed by atmospheric researchers might be produced readily under conditions stimulating release of such particles.

The production of atmospheric ice nuclei by decomposing vegetation was studied by Schnell and Vali (1972). The authors postulated that leaf-derived nuclei (LDN) might have a use as a cloud seeding agent, although their efficiency is much lower than that of silver iodide. The best-characterized active component of LDN has since been found to be *Pseudomonas syringae*. Although this has not been further developed for use as a nucleating agent, research is continuing on the role of these organisms in plant sensitivity to freezing, a problem of widespread economic concern.

8.9 CONCLUSIONS

Silver concentrations in plants vary with the season, plant species, silver form, and ability of the plant to accommodate the presence of environmental silver.

Laboratory (greenhouse) studies show that silver from silver iodide is taken up to a much lesser extent than is silver ion from nitrate. With increased silver levels in soils, increases in aboveground parts are observable; but on a relative basis, less material will be translocated.

Silver tends to accumulate mainly in the root zone, and based on silver monitoring which has been carried out in conjunction with several Colorado projects, no changes in environmental silver levels have been detected which can be related to these programs.

The use of trees as metal monitoring points has been studied and several investigators have evaluated silver uptake characteristics, typical background levels, and geological effects on silver uptake by trees.

The physiological effects of silver nitrate have been well studied, but it is difficult to predict potential effects of silver iodide based on this information, as silver ion has been used in most studies. Silver iodide, at levels much higher than may be expected under seeding conditions, has been shown to inhibit germination and growth of various seedlings, and to reduce pollen tube length. At lower levels, silver iodide has shown either no effects or has shown stimulation of specific plant growth processes.

Levels of silver in food crops have not been studied in detail, but wheat milling products, corn, and other products contain silver, and these may be useful for monitoring the metals exported from agricultural areas which would be impacted by weather modification.

The phenomenòn of metal-accommodation in plants makes it difficult to predict silver impacts. Silver will tend to accumulate in the plant root zone, probably in the elemental form, and will thus have negligible effects on plant functioning.

Silver and other metals may influence the microflora on plant surfaces, which might be useful for monitoring possible silver accumulation from precipitation. In addition, the release by plants of microparticles containing metals, including silver, to the atmosphere, may lead to secondary nucleation phenomena.

In summary, available field and laboratory studies, using silver iodide at levels much higher than might be expected to occur as a result of weather modification, indicate that the accumulation of this type of nucleating agent in surface soil zones should have no discernable effects on plant growth and physiological processes.

9

Potential Impacts on Aquatic Systems*

D. A. Klein

9.1 INTRODUCTION

The movement, cycling, and final deposition of metals in aquatic systems has been of concern for many years, as metals can concentrate at specific sites through biotic and abiotic processes, and because organisms may come in more direct contact with metals in these environments. Previous general assessments of environmental exposure (Smith and Carson, 1977) and of potential impacts of nucleating agents (Cooper and Jolly, 1969, 1970) also suggest that if silver has environmental impacts, these would probably first be observed in aquatic environments, due to lower organisms having intimate contact with metals dissolved in their environment, and to their relative organizational simplicity.

Stumm and Bilinski (1973) discussed the analysis and interpretation of data on metals in aquatic environments. Trace metals in waters do not behave completely as chemical equilibria might predict, due to the presence of many complex organic and inorganic compounds which may influence biological activity and metal ion transport. The flux of silver and other metals has also been found to exceed the "natural" rate of cycling, resulting in increased metal availability. These authors stated that our ignorance of the speciation of trace metals makes understanding their functions difficult. For example, the solubility of Ag_2S depends on the presence of $AgSH$, $Ag(SH)_2$, and $Ag_2S_3H_2^{-2}$, but information about these complexes is scarce. Additional information regarding this area is provided in Chapter 4. Also the complexes tend to adsorb more at interfaces and surfaces than do metal ions. They emphasized that organic substances in water may reduce free metal ion concentrations, making them less avail-

*Major contributions to the completion of this chapter were made by Gayle Robison.

able to interact with biological systems, even though total soluble metal concentrations (in various low-weight organic complexes) may be relatively high.

Silver radiotracers have been used in many studies of these processes. The adsorption-desorption equilibria of ^{110}Ag were considered by Murray and Murray (1973) in an informative study of physical parameter effects on silver fluxes in marine environments. At higher pH values, both adsorption and desorption of silver by river sediment increased (Fig. 9-1). Increasing sediment load also increased the total adsorption capacity for a given metal, but proportionally, each increment increase of sediment gave a lower response. Desorption appeared to decrease 10% at very high sediment levels. The authors concluded that since pH, sediment, and metal levels have significant effects on adsorption and desorption, these factors may be used to estimate the amount of silver which can be taken up by the system. Under static conditions, adsorption and desorption proceeded at markedly different rates (Fig. 9-2), with adsorption to sediment materials

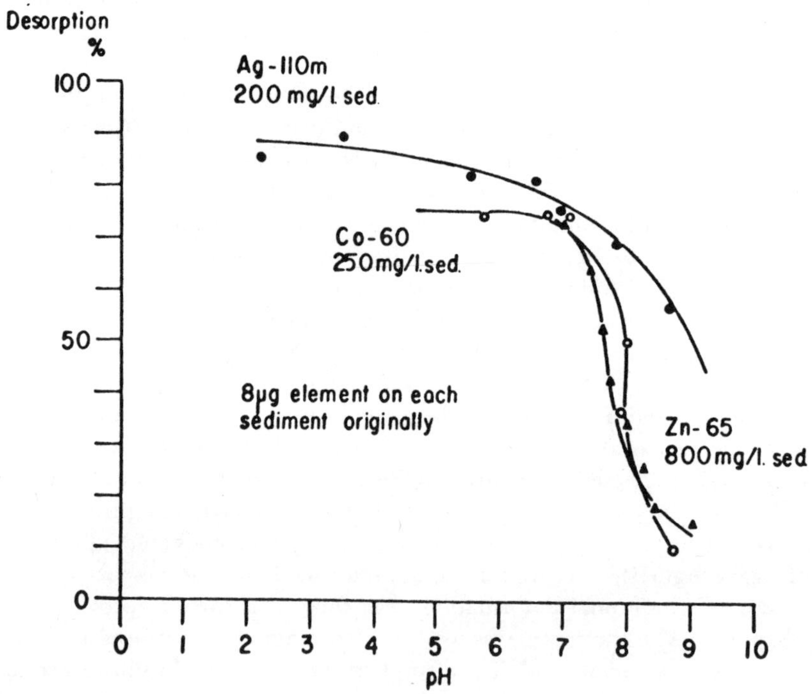

FIGURE 9-1. *Variation of desorption of trace elements at constant adsorbed trace-element concentration and sediment load with varying pH (Murray and Murray, 1972).*

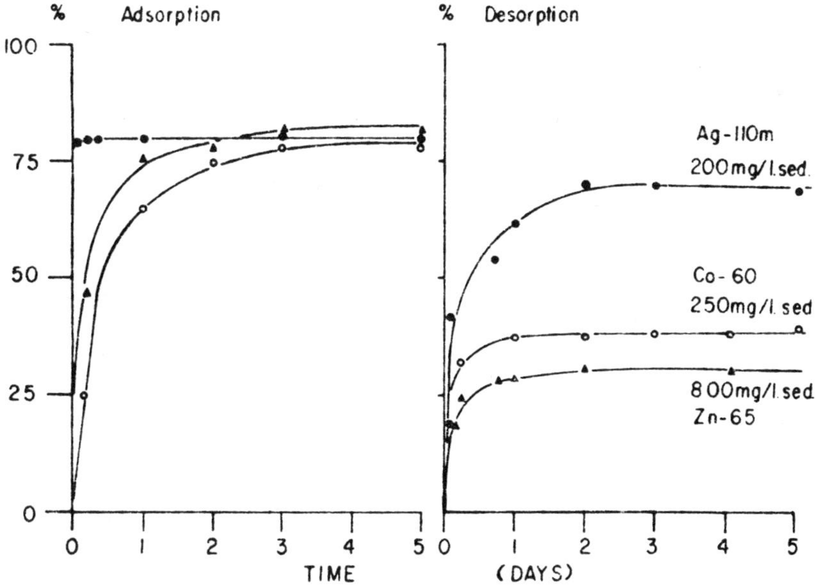

FIGURE 9-2. *Equilibrium time for adsorption and desorption processes under constant pH conditions (Murray and Murray, 1972).*

proceeding at a much faster rate. The adsorption process was complete in less than 5 to 6 hours, while the corresponding silver release from sediment to a nonsilver containing solution required approximately 3 days to reach equilibrium.

Other silver complexes may play lesser but nonetheless significant roles in aquatic systems. Feitknecht and Schindler (1963) briefly discussed silver oxides and their possible role in metal concentration processes in sediments. Precipitation of Ag^+ solutions apparently yields Ag_20, and several hydroxy complexes (AgOH, $Ag(OH)_3^{-2}$) were observed. No information was given on the levels of these compounds in natural waters.

9.2 MONITORING OF ATMOSPHERIC FLUXES TO AQUATIC ENVIRONMENTS, AND OF FOOD CHAIN BIOCONCENTRATION PROCESSES

Heavy metal impacts on oceans have not been studied to any great extent. However, Duce et al. (1972) found high metal concentrations in the surface microlayer (top 100 to 150 μm) of mud from Narranganset Bay, Rhode Island. They proposed that organic substances in the surface mud layers may provide complexing sites for many heavy metals, resulting in increased transportation and concentration. The author noted that pollutants in the surface microlayer can enter the food chain, but conceded that effects of many of these metals on aquatic organisms are largely unknown. As the surface microlayers of most muds will be oxidized, this may lead to an increased ability of metals to be mobilized through chemical oxidation processes. In research concerned with radionuclide sedimentation (Simpson, 1976), the accumulation rates for silver and other metals in the Hudson River estuary were examined, and most radionuclides were confined to the top 5 to 10 cm of sediment, as might be expected.

Hodge et al. (1978) studied the influence of atmospherically-transported aerosols (including silver) on the composition of surface ocean waters. It was known that metals in the solid phase can leach into sea water, but the contribution of aerosols to heavy metal levels in oceans was not known. The authors observed high levels of some elements in the winter months, and associated these with atmospheric phenomena. The residence time (airborne) of 0.1 to 1.0 days for silver indicated short range transport of large particles. It was found that aerosol input may supply up to 25% of the silver in coastal surface waters of southern California, although these fluxes do not account for the accumulation of this metal in coastal sediments. Leachable silver levels in the coastal waters averaged 0.04 ng/m^3, and ranged from 0.01 to 0.13 ng/m^3. Total average silver levels were approximately 0.05 ng/m^3 at La Jolla and 0.03 ng/m^3 at Ensenada, suggesting that combined natural and anthropogenic sources of silver may reach levels which could interfere with the detection of added nucleating agents.

Bruland et al. (1974), in a related study, determined the fluxes of metals, including silver, into sediments off the coast of southern California. Silver accumulated in localized regions, perhaps having been mobilized by other materials released by man. As a possible source of silver in this environment, waste water from Los Angeles County transports 15 tons of silver per year to coastal waters.

The distribution of heavy metals in sediments of a large Norwegian fjord was examined by Skei et al. (1972), where three industrial plants discharged up to 30 kg of silver per day. Sediment silver levels ranged from 13 to 190 $\mu g/g$, averaging mostly under 50 $\mu g/g$. Good positive correlations existed between silver and copper levels, as noted in Fig. 9-3. The authors also cited Krauskopf (1956), who found silver levels of 0.1 $\mu g/g$ in naturally occurring marine shale sediments. Since certain metals are selectively fixed onto particle surfaces, Skei and his co-workers postulated that organic matter settling to the bottom affected metal concentration processes in these sediments.

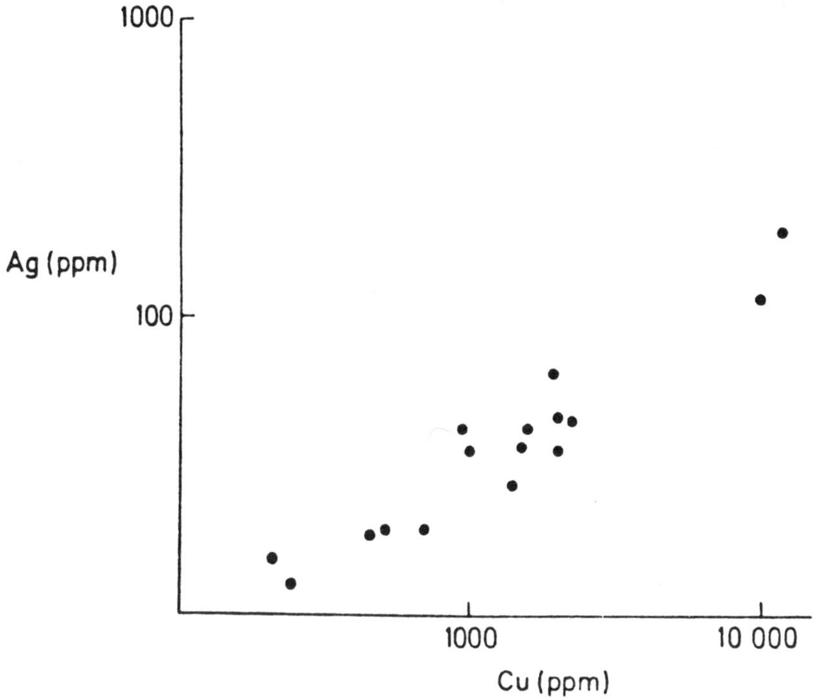

FIGURE 9-3. *Correlation of Ag and Cu in surface sediments of Sorfjord, Norway (Skei et al., 1972).*

The close correlation found between copper and silver levels in marine sediments is important, due to the known increased toxicity to algae of copper ion-silver ion combinations (Young and Lisk, 1972), and to the proposed use of silver iodide-copper iodide mixtures in weather modification programs (Passarelli et al., 1973). The copper apparently permits a better crystal lattice fit, resulting in threshold nucleation temperatures as high as -0.5°C for a precipitate of CuI-AgI. Once again, the addition of copper to aquatic systems in combination with silver could be of environmental interest, if this were carried out on a large scale.

The possibility of silver bioconcentration in higher trophic levels of aquatic ecosystems has been analyzed by measuring metal-calcium ratios. Using samples from coral reefs, Veeh and Turekian (1968) found that silver levels ranged from 0.014 μg/g in Samoan coral to 0.100 μg/g in Hawaiian coral. The generally low Ag:Ca ratios indicated discrimination against silver relative to calcium, compared to levels found in sea water. Lead also was relatively excluded from the higher trophic levels of a lake aquatic system in the California sierras, according to the metal-calcium ratio (Bradford et al., 1968).

9.3 FRESH WATER ENVIRONMENTS—DEPOSITION PROCESSES

Numerous studies of trace metal levels in freshwater lakes, streams, and rivers have been conducted. Kleinkopf (1960) utilized spectrographic analyses to determine the trace element content of lakes in northern Maine, and analyses showed that one random sample per lake would be an accurate gauge of the overall mineral content. Silver levels were generally quite low, with 330 of 439 samples containing less than 0.5 μg/liter of silver. The author noted that each lake appeared to be an indicator of the average metal levels of the surrounding area.

Kopp and Kroner (1970), in a five year summary of trace metals in U.S. lakes and rivers, noted that the permissible silver level for public water supplies was 50 μg/liter, but felt that a desirable level would be no silver at all. Of 1,577 samples taken throughout the United States, 6.6% contained detectable silver, with concentrations ranging from 0.1 to 38 μg/liter and having a mean of 2.6 μg/liter.

Bradford et al. (1968) analyzed the silver content of water from 170 high Sierra lakes in California. Silver was detected in 146 of the 170 lakes tested, at concentrations ranging from 0.03 to 6.0 μg/liter, with a median value of 0.1 μg/liter. These workers noted that optimum phytoplankton growth depends on the levels of trace and major elements, and trace elements effects, including those of silver, were felt to be more pronounced in waters with a low mineral content.

Kharkar et al. (1968) analyzed water samples from 12 rivers to estimate metal types and amounts that reached the ocean in a dissolved state. Four possible modes of metal transport were noted: (1) in solution, (2) in the lattice of minerals of the suspended sediment load, (3) adsorbed onto minerals, and (4) associated with organic material. Silver had a relatively narrow range of concentrations in the rivers studied, with a high average of 0.55 μg/liter in the Klamath and a low of 0.10 μg/liter in the Russian River, both in California. No relationship was observed between the dissolved and adsorbed metal concentrations, and less than 10% of the silver burden was in a desorbable form when it later came in contact with sea water (Table 9-1).

Turekian and Scott (1967) examined concentrations of trace metals, including silver, in suspended material in streams. Regional differences in trace metal levels were considered to be due to industrial and domestic

TABLE 9-1. *Dissolved, desorbable and total supply of silver to the oceans from selected rivers (Kharkar et al., 1967).*

River System	Dissolved in Stream (μg/liter)	Adsorbed in Stream and Desorbed in Contact with Sea Water (μg/liter)	Total Soluble Load (μg/liter)
Mississippi	0.24	0.02	0.26
Susquehanna	0.37	0.02	0.39
Mad	0.26	0.02	0.28
Neuse	0.36	0.02	0.38
Rhône	0.28	0.03	0.41
Amazon	0.23	0.01	0.24

sources, organically associated metals, or the presence of minerals rich in trace elements. The highest silver levels, 15 μg/g, were found in suspended materials from the Susquehanna River, Pennsylvania, where 4.5 tons of silver were estimated to be released per year (based on a recalculation from their data).

The role of aquatic sediments in immobilizing and concentrating silver in overlying waters has also been emphasized in other studies. Laskowski et al. (1975) analyzed the heavy metal content of recent sediments near Mainz. The water channel from which grab samples were taken was known to be heavily contaminated with metals, and silver levels ranged from 14 to 180 μg/g on a dry weight basis, with an average of 63 μg/g for 14 samples. Fjerdingstad et al. (1976) analyzed heavy metals (not including silver) in sediments from Danish lignite pits. In the bottom sludges, metal concentrations ranged from 10 to 100 μg/g, which indicated the tendency of sediments to concentrate and retain heavy metals. Interestingly, the authors found no correlation between heavy metal levels and changes in numbers of *Thiobacillus* organisms.

Two Idaho studies (Maxfield et al., 1974a,b) discussed the heavy metal content of sediments from the Coeur D'Alene Lake and river delta, where heavy metal pollution occurred due to mining. Silver was present at an average of 10 μg/g, while copper and cadmium levels were 4 to 10 times higher.

Extensive sampling of stream sediments in the Uncompaghre and San Juan Primitive areas (U.S. Geological Survey Bulletin, 1968) has shown relatively low silver levels compared to other sediments studied, which is interesting in light of the natural geology, long-term mining activities, and weather modification programs which have been carried out in this area.

Klein and Striffler (1976) determined silver concentration in aquatic systems in the area impacted by the National Hail Research Experiment.

Sediments and plants contained consistently less than 0.2 μg/g silver (Table 9-2). Waters contained an average of 0.014 μg/liter of silver in 1971, in spite of four years of hail suppression activity. All monitoring sites gave very low bottom sediment silver concentrations.

9.4 BIOLOGICAL CONTRIBUTIONS TO SILVER DYNAMICS IN AQUATIC ECOSYSTEMS

Much available information indicates that marked increases in silver concentrations can occur in localized environments in aquatic ecosystems, through biological as well as physical-chemical processes. However, as reduction to less active metallic forms as well as complexing with organic and inorganic materials can be involved, such concentration *per se* may not directly relate to increased toxicity.

Silver levels and concentration factors in aquatic organisms vary greatly from species to species. Noddack and Noddack (1939) found a concentration factor of 22,000 over sea water silver levels in aquatic organisms. Vinogradov (1953) discussed early qualitative and quantitative assessments of silver in marine organisms, and noted that echinoderms without a well-developed liver (excepting *Asteroidea*) did not accumulate heavy metals. Aquatic invertebrates, Vinogradov added, concentrate metals such as iron, zinc, manganese, cobalt, vanadium, and copper which are used in metabolic processes, but silver was not specifically noted as having any biochemical function.

Trace metal levels in beach-dwelling amphipods (beach hoppers) and dipterans (marine flies) were determined by Bender (1975). In spite of these insects feeding in an area where silver might be available from sedi-

TABLE 9-2. *Mean silver concentrations in terrestrial and aquatic components, 1971 to 1974 (Klein and Striffler, 1976).*

Ecosystem Component	Mean Silver Concentration (μg/g ash)				Analysis of Variance (One Way Test)
	1971	1972	1973	1974	
Terrestrial					
Soil	0.014	0.014	0.032	0.02	Significant at 1% level
Grass	0.145	0.091	0.098	0.055	Not Significant
Aquatic					
Sediment	0.017	0.036	0.034	0.017	Not Significant
Plants	0.03	0.074	0.154	0.079	Not Significant
Water (mg/liter × 10^{-5})	1.4	0.650	0.560	0.092	Significant at 1% level

ments and waters, they had silver levels ranging from undetectable to less than 3 μg/g on a dry weight basis. Bender suggested that marine contamination could impact on the terrestrial ecosystem via the beach interface.

In an extremely informative study, Cowgill (1973) examined biogeochemical cycles of many elements, including silver, in water lilies and their aphids in Linsley Pond, Connecticut. The author initially postulated that the aphids might bioconcentrate silver contained in the water lilies, however, the aquatic plants contained from 0.06 to 2.3 μg/g silver (with an average of about 0.5 μg/g) while the aphids averaged only 0.42 μg/g silver (Table 9-3). The plants showed only a two-fold increase in silver levels over the silver levels in the substrate, and contained about the same amount of silver as deep water sediments. Of special interest was the extremely low silver content of waters which left the lake system, 0.00012 μg/liter, with copper being present at a 100-fold higher concentration. This was only slightly higher than the average silver concentration noted for 400 main lakes (Kleinkopf, 1960). Generally, silver was felt to be incapable of diffusing from the reduced mud surface in the presence of FeS, unlike mercury, which could form stable, volatile methyl compounds. The author also suggested that silver enrichment of lake water may be due to increased silver iodide use in cloud seeding. Generally the Cu:Ag ratio was much lower in the plants and insects examined than in the earth's crust.

Cowgill and Burns (1975) also measured the metal content of two *Daphnia* species growing on *Euglena* or mixed fresh water algae, to determine if bioconcentration was occurring from spring water via the algae. The mixed algae concentrated silver 7.5-fold over the levels in water, but the *Daphnia* did not have higher silver concentrations, even when fed a steady diet of these algae.

Black and Mitchell (1952) also analyzed marine algal specimens for trace metal levels, to determine if concentration might be occurring. Silver was present at very low levels, with *Laminaria* consistently having 0.7 μg/g of silver or less. Silver levels were slightly higher during the winter months, with a distinct decrease being noted between May and June. Deep sea water in the area contained 2.9 μg/liter of silver, indicating that the element did not concentrate to any great extent in algae, unlike metals such as titanium, zinc, and nickel.

Luoma and Jenne (1975) studied the availability of sediment-bound silver to a deposit-feeding clam and found it to be dependent on the physical and chemical nature of the metal-sediment association. Silver was taken up from organic detritus and iron oxide precipitates via surface exchange, complexation, and chelation. It was noted, however, that other relatively minor sinks in sediments may be important metal sources for the clam, *Macoma baltica*, as silver uptake from carbohydrates and calcites was even greater than from organics and oxides. The authors emphasized

TABLE 9-3. *Copper and silver levels in Linsley Pond, in comparison with data from other sources (Cowgill, 1973).*

Locality	Metal Concentration, μg/g Cu	Ag
Linsley Pond		
Soils	105.0	0.02
Rocks	94.5	0.01
Sediments		
Deep water	150.0	0.52
N. odorata	192.8	0.23
Outlet water	0.012	0.000012
R. nymphaeae	36.3	0.42
N. odorata (1971)		
Flowers	45.3	0.66
Flower stalks	44.5	0.30
Leaves	49.8	0.32
Stems	39.5	0.30
N. odorata (1972)		
Flowers	35.6	0.56
Flower stalks	44.2	0.40
Leaves	36.7	0.60
Stems	37.7	0.56
Cedar Lake		
Water	0.013	0.000011
N. odorata (1971)		
Flowers	46.5	0.94
Flower stalks	45.0	0.24
Leaves	52.1	0.26
Stems	45.7	0.20
Kleinkopf (1955), 440 Main Lakes	0.010	0.000094
Bowen (1966)		
Land plants	14.0	0.06
Insects	50.0	≥0.07
Taylor (1964), Earth's Crust	55.0	0.07

the importance of considering the geochemical characteristics of aquatic sediments when predicting the susceptibility of bottom-dwelling fauna to silver contamination.

Luoma and Jenne (1976) developed correlations between the bioavailability to clams of silver complexed with sediment components (organic detritus, iron oxides, organically-coated iron oxides, manganese oxides, and inorganic and biogenic carbonates) and chemical extractability. Silver had a high bioavailability from all types of sediments tested, and uptake by the clam correlated well with most parameters of noncrystalline silver. In contrast, the uptake of cadmium, cobalt, or zinc was strongly related to physical form in the sediment. Weak acids, reducing agents, and oxidizing agents removed significant quantities of the four metals from the various physiochemical sinks without being particularly selective.

Uptake of silver by certain New Zealand bivalves has also been considered. Brooks and Rumsby (1965), measuring silver content on a dry weight basis in the soft portions of whole shellfish, found levels of less than 0.1 μg/g in mussels, 5.6 μg/g in oysters, and 0.7 μg/g in scallops, and efforts were made to determine the uptake mechanisms. The internal organs consistently yielded higher silver levels than muscle tissue and shell material. A distinct relationship between copper and silver was also found. Where particulate absorption occurred, copper levels were higher than usual; and with ingestion of algae and small amounts of sediment, silver levels were higher, especially in the heart and visceral mass of the oyster, the visceral mass of the scallop, and at much lower concentrations in mussel organs. Compared with concentrations in the marine environment, the enrichment factors for silver in the soft parts of the animals were 2,300, 18,700, and 330 for scallop, oyster, and mussel, respectively.

Bertine and Goldberg (1972) conducted a similar study with clams, mussels, and shrimp to determine man's influence on the trace metal content of these marine organisms. For this purpose, both museum and fresh specimens were analyzed. Silver levels in clams showed no significant increase, and in mussels the levels actually decreased from 0.070 μg/g to 0.009 μg/g on a dry weight basis from 1870 to 1970.

An interesting test system which evaluates the uptake and concentration of metals from sediments has been carried out with annelid worms (polychaetes), predominantly *Nereis diversicolor*. Bryan (1975) found a relationship between the general toxicity of a metal (in its ionic form) and its ability to be concentrated above its background concentration in sea water (Fig. 9-4). Silver (as the free ion) concentrated approximately 100-fold on a dry weight basis over its concentration in the sea water medium, and similar relationships were found for copper and mercury. The regulation of silver absorption was poor (Fig. 9-5), as the silver concentration in the sediments and *Nereis* were similar, while for several other metals (lead and zinc especially), some degree of regulation was suggested. In earlier studies, Bryan (1974) found that scallops could accumulate silver (along with cadmium and copper), and that granules could occur in the digestive tract. Also, the test organisms may have developed resistance to high levels of silver and other metals to which they have been exposed. An interesting point revealed by these studies was the increased silver absorption by population adapted to a high lead and high copper environment. These results suggested that the excess metal-binding capacity of the epidermal cells was being titrated with silver. Again, as observed by earlier workers, there was less difference in silver uptake between high- and low-copper accommodated populations, in spite of the chemical similarity between these two metals. The author noted that development of tolerance will even-

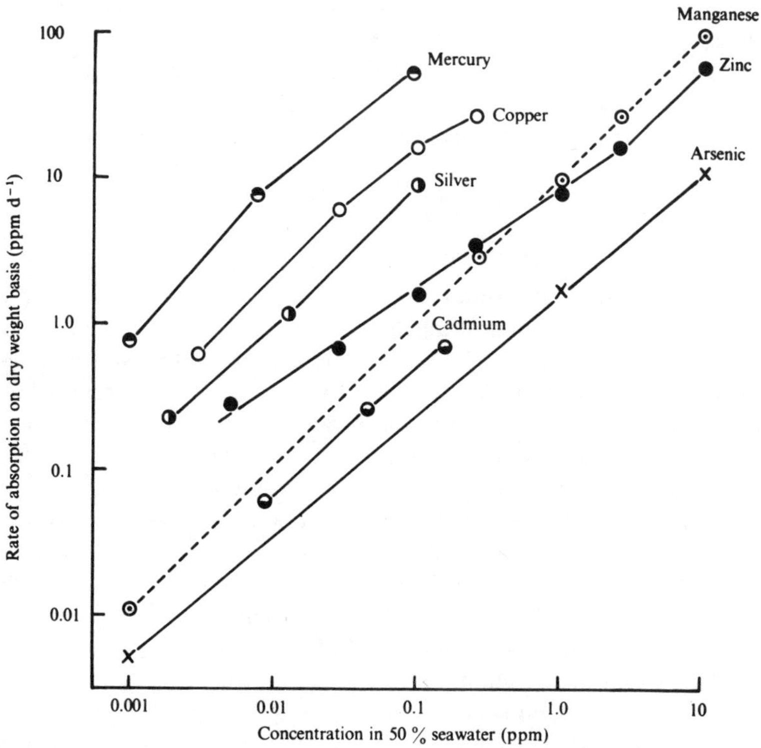

FIGURE 9-4. *Rates of absorption of metals from different concentrations in 50% sea water by* Nereis diversicolor *from Avon estuary (Devon). Each point is a mean value for five worms and measurements were made using radioisotopes at 13°C. Mercury was added as chloride, silver as nitrate, arsenic as arsenate, and others as sulphates (Bryan, 1975).*

tually lead to a normal distribution and diversity of species even in areas of heavy contamination, a phenomenon which will provide a challenge to individuals and organizations involved in the preparation of environmental impact statements.

Lande (1977) recorded effects of heavy metal pollution on plant and animal life in a Norwegian fjord. Mussels contained 1 to 6 μg/g silver in the soft parts, and 2 to 4 μg/g in the shell. Limpet tissue levels were lower, but the shells contained 4 to 5 μg/g. All fish examined had from 1 to 2 μg/g silver, and gulls averaged 2 μg/g. Eider livers appeared to concentrate the metal, with a concentration of 44 μg/g. All of the dominant species found in the study were tolerant to variations in environmental factors, including metal exposure.

9.5 POSSIBLE EFFECTS ON AQUATIC MICROORGANISMS

The available literature on potential effects of silver ion and silver iodide derived from nucleating agents on heterotrophic microorganisms is summarized in Chapter 6. For low levels of silver iodide in waters, effects on typical heterotrophic microorganisms are expected to be negligible, especially if organic matter, clay, or halides might be present to react with and inactivate silver ions. Additional information is available on the possible effects of various silver forms on phyloplankton and aquatic primary productivity. Much of this work relates to developing procedures for the control of algal growth in reservoirs, which might be caused by nutrient enrichment from detergents, sewage treatment plants, and agriculture. Copper (as the sulfate salt) has been the major metal used to control algal growth, although silver use has been considered.

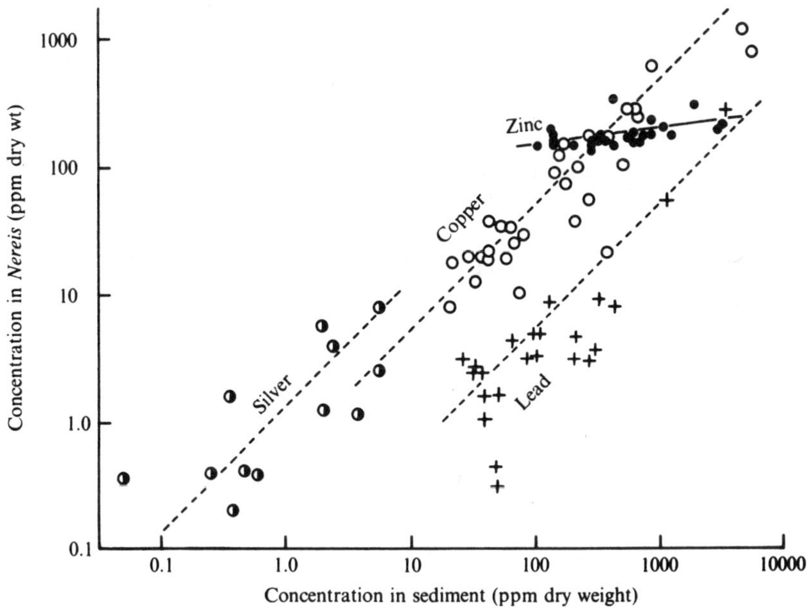

FIGURE 9-5. *Relationship between concentrations of metals in* Nereis diversicolor *and those in sediments from more than 20 estuaries in Devon and Cornwall. Each point is a mean value from an individual estuary or part of an estuary. Broken lines show direct proportionality (Bryan, 1975).*

Fitzgerald et al. (1952) studied the possible effects of chemicals with selective toxicity to blue-green algae. Various organic and inorganic substances were used, but no mention was made of silver. The authors did note, however, that copper sulfate used as an algicide had no effect on invertebrate larvae in lakes where the compound was used. Other studies of copper use as an algicide and or more general aspects of algistatic and algicidal agents have been completed by Fitzgerald and Faust (1963) and McIntosh (1974, 1975). These are important in that they provide information about a metal that may have effects similar to those of silver on various components of the biota.

Young and Lisk (1972) conducted experiments to determine if a combination of copper and silver ions would be more effective in controlling algal blooms, while being less toxic to fish and other wildlife, than copper alone. A level of 0.3 mg/liter of copper-silver mixture was lethal to all blue-green algae tested, but green algae showed resistance to the mixture. Copper and silver, in conjunction, were slightly more effective than copper alone, but the authors noted that possible toxicity to fish and increased costs might outweigh any benefits gained by their use.

Fitzgerald (1967) investigated the possible algistatic properties of silver, and various factors affecting its toxicity. *Chlorella* growth was prevented by 7.5 mg/liter of silver nitrate, while 120 mg/liter was required to kill the algae. The presence of thioglycollate negated the toxic effect of silver ion, as would be expected. In addition, silver which had been adsorbed onto vessel surfaces did not inhibit growth of this algae. In the presence of 12 mg/liter of sodium iodide, 60 mg/liter of silver nitrate caused no *Chlorella* inhibition.

Studies carried out at Colorado State University using silver iodide from seeding agent burn complexes suggest that with high levels (1.06 mg/100 ml) of silver iodide, transient effects on cell yields and chlorophyll production can be observed (D. A. Klein and N. Nagle, personal communication). Using a *Scenedesmus* species, distinct inhibition of chlorophyll a production occurred after 28 days of incubation, without corresponding decreases in biomass being noted. By the 40 day assay, the populations were approximately equal, based on optical density measurements; and the chlorophyll a production was equal to that of controls to which plain glass filter disks had been added (Fig. 9-6). With the presence of the AgI-TiO_2 complex, a distinct increase in chlorophyll a accumulation was noted.

Increasing emphasis is being given to the use of algae in assessing water quality, which may provide methodology useful for evaluating possible nucleating agent impacts. Patrick (1973) described saprobic indexes, with which it is possible to monitor species diversity and community structure changes in relation to system perturbations. The effects of organic versus "toxic" pollutants, the latter possibly including various

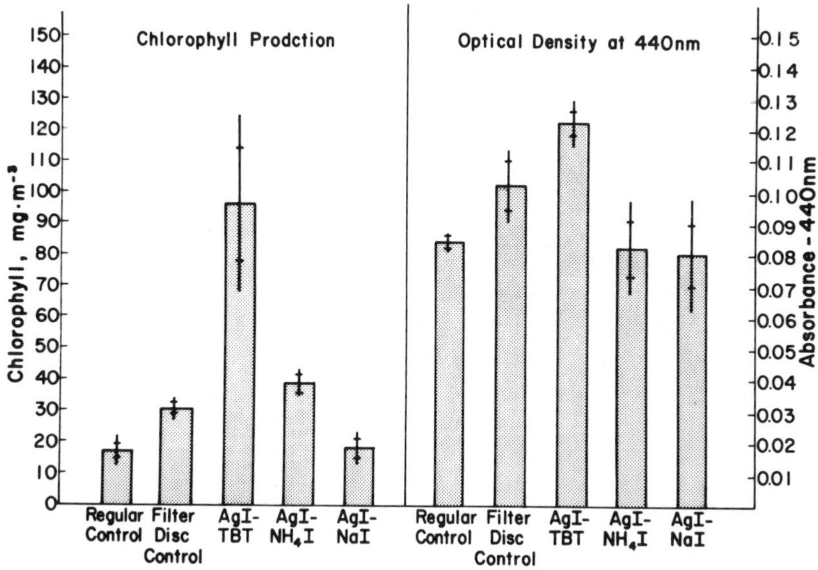

FIGURE 9-6. *Effect of AgI burn complexes on chlorophyll production and optical density of* Scenedesmis *(40 day incubation). The burn complexes used added at approximately 1.04* μg per 150 ml of culture solution.

silver forms, were felt to be quite different. Since algal species vary in their range of tolerance to particular pollutants, it may be necessary to analyze community structure changes. Measurements of chlorophyll or biomass alone might not allow the effects of such materials to be detected.

9.6 EFFECTS ON HIGHER AQUATIC ORGANISMS

As noted for other areas, most studies of silver toxicity to higher aquatic organisms have used silver ion, perhaps due to the solubility and to the known effects of this very reactive silver species. Unfortunately, this information is of little relevance in predicting environmental impacts of nucleating agents, unless precise information on the chemical composition of the water is provided. With such information, the silver ion in a particular environment could be calculated. Typical research using silver ion to assess the potential "biological effects of metals" has been carried out using clams and other marine bivalves. From a toxicity as well as an accumulation standpoint, clams and aquatic invertebrates are important as a first consumer level. These organisms function in bottom sediment and beach environments where heavy metals (including silver) accumulate at much higher levels than are found in the overlying waters (Bryan, 1975), and where concentrations are up to 8 to 10 times higher than those of underlying sediments.

Much of the work on clams and other bivalves has been concerned with the possible effects of metals on the gametes, embryos, and fry, as these can be more sensitive to metal presence in the environment (Doudoroff and Katz, 1953; Calabrese et al., 1973). As noted by Bryan (1975), the mechanism by which organisms handle natural fluctuations in the availability of heavy metals can be particularly important under contaminated conditions.

Calabrese et al. (1977a) studied clams, mollusks, and scallops, and observed typical toxic effects (LC_{50}) at silver ion levels of 25 μg/liter for larvae of *Crassostrea virginica* and *Mercenaria mercinaria*. These workers did, however, indicate that their data applied only to the inorganic salts tested, and "did not take into account the form of the metals when in solution or the background levels of the metal in the test medium," which for silver was 2.8 μg/liter. Similar work has been completed by Calabrese et al. (1977b), Calabrese and Nelson (1974), and Thurberg and Collier (1977). Without detailed information on the chemical composition of the water, or without calculations or measurements of the free silver ion concentration, it is difficult to determine the significance of these studies.

Other studies by these workers revealed some positive effects of silver ion presence. Thurberg et al. (1974) found that hard clams increased their oxygen uptake when decreasing salinity and increasing silver ion concentration occurred (Fig. 9-7). Similar effects were observed with mussels. Soft-shelled clams showed a marked increase in oxygen use at low salinity and 1 μg/g silver, but all died when exposed to higher salinity and 1 μg/g silver. Copper exhibited the opposite effect. The authors stated that further studies are necessary in order to determine long-term and chronic effects of such metal ions in these environments.

In any case, growth retardation by metals would prolong the period of time during which the embryo is susceptible to predation and disease. The differences in test conditions used in these studies make it difficult to draw general conclusions, but with the low silver ion levels which could be expected, stimulation as well as inhibition of aquatic microbiological processes might occur.

Much additional work is available showing the sensitivity of lower invertebrates to silver ion, usually in the range of 0.05 to 0.10 mg/liter. Snails were tested by Harry and Aldrich (1958). After a period of high water flow into Puerto Rican streams, the heavy metal content of the sediments ranged from 0 to 40 μg/g. Generally, the snails were found only in streams which had maximum metal (primarily copper) concentrations not exceeding 10 μg/g, predominantly derived from natural ores.

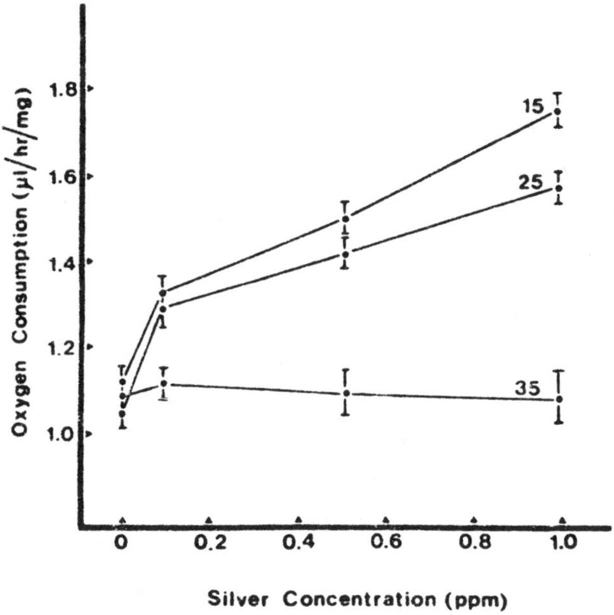

FIGURE 9-7. Crassostrea virginica. *Effects of silver on oxygen consumption. Each point comprises the mean gill tissue oxygen consumption value of 16 animals. Bars: standard errors. Each line represents a series of experiments conducted at the salinity shown thereon (Thurberg et al., 1974).*

A study involving planaria also provided information on the responses of a lower invertebrate to the presence of silver ion, and on relationships between ionic metal solution pressures and toxicity (Jones, 1941). In this study, silver ion had a threshold toxicity of 0.15 mg/liter (1.4×10^{-6} M) at pH 6.6. In calculating ionic "solution pressures" against threshold toxicity levels, interesting relationships were observed (Fig. 9-8). The grouping of silver, copper, and mercury was again evident.

Recently Nehring (1973) studied the susceptibility of aquatic insects to silver, copper, lead, and zinc.* This work was carried out to monitor fish kills "after the fact," using insects as a bioassay organism. The acute toxicity of these metals to two species of aquatic insects was compared with results for fish.

*This section contributed by P. H. Davies and J. P. Goettl.

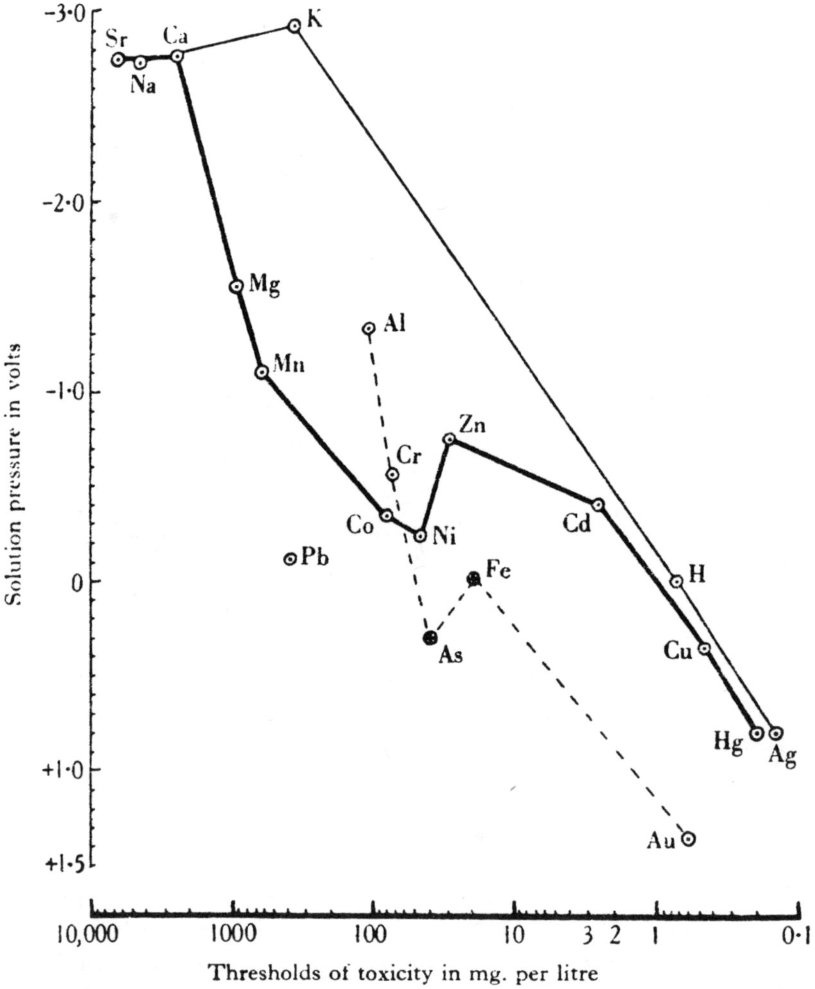

FIGURE 9-8. *The relation between the solution pressures of the ions and their thresholds of toxicity. Explanation in text. Metals whose position on the solution pressure scale is uncertain, or whose threshold of toxicity cannot be definitely determined, are indicated by crossed circles (Jones, 1941).*

The tests were conducted by exposing *Ephemerella grandis* (mayfly), *Pteronarcys californica* (stonefly), and *Claassenia sabulosa* (predaceous stonefly) to silver from silver nitrate in the same manner used in flow-through fish toxicity tests. Ten stonefly naiads were used in each exposure concentration while tests with mayflies had 50 organisms per silver concentration. A control containing the same number of organisms as the toxicant concentration was used in each of the tests. The tests were checked for deaths five times daily. Dead naiads were removed, rinsed in distilled water, and dried to a constant weight. The dry insects were digested in nitric acid and analyzed for silver by flame atomic absorption spectrophotometry (Adrian, 1971). Table 9-4 summarizes the results of six short-term toxicity tests. An LC_{50} value was established only in test number three. This 15-day value was 8.8 μg/liter. An LC_{50} value could not be established in the other tests because mortality in the lowest concentration exceeded 50%.

The conclusions that can be drawn from this study are that small *P. californica* are most susceptible to silver toxicity than are later, larger instars, and that males are more likely to die than females. Nehring (1973) showed that, based on time of death, *P. californica* are more resistant to silver effects than *C. sabulosa* or *E. grandis*. He observed that the most likely time of death for immature aquatic insects was during a molt. Expansions of mortality curves place the long-term "no effect" concentration for silver on aquatic insects below 1 μg/liter. Silver concentrations in the aquatic insects tested demonstrated a high correlation coefficient with the environmental silver concentration, a relationship which could be valuable for indirectly calculating silver concentrations in water.

Some information is available on the possible effects of silver on aquatic plants. The effects of some SH-function inhibitors on *Lemna perpusilla* (duckweed) were examined by Takimoto and Tanaka (1973). It

TABLE 9-4. *Results of six short-term toxicity tests with aquatic insects and silver (Nehring, 1973).*

Test no.	Species	$\bar{x}$ length (mm)	Exposure time (days)	Lowest Ag ion concentration (μg/liter)	% mortality	r[1]
1	*P. californica*	37	6	50	60	0.99
2	*C. sabulosa*	—	3	50	86	0.95
3	*P. californica*	40	15	6	30	0.93
4	*P. californica*	24	7	4	80	0.90
5	*E. grandis*	—	7	60	100	0.89
6	*E. grandis*	—	10	9	96	—[2]

[1]Correlation coefficient of silver concentration in the test water versus silver concentrations in the test insects.

[2]Samples destroyed.

was shown that EDTA nullified day length-independent flowering behavior caused by copper. Silver ion added to the growth medium increased long-day flowering, but not as much as copper or potassium.

9.7 LEVELS OF SILVER IN DRINKING WATER COMPARED TO LEVELS FOUND IN NATURAL WATERS

With the natural propensity of silver to adsorb to organic materials and to accumulate in bottom sediments, silver levels in effluent waters will generally be much lower than in influent waters (Cowgill, 1973).

McCabe (1969), of the U.S. Bureau of Hygiene, conducted a survey of trace metal content of drinking water from the Chicago area. At each treatment plant silver levels were determined. No silver level exceeded the Public Health Service standard of 50 μg/liter. The highest level found was 8 μg/liter in a portion of the distribution system. Four hundred and ninety-six of the 550 samples had 3 μg/liter or less.

Bradford (1971) discussed the significance of trace metals in five areas of water use: recreation, public water supplies, aquatic life, agriculture, and industry. Water samples from various areas of California were examined, and of 383 samples, only 21 had silver levels above 3 μg/liter. Most drainage and waste waters contained silver, whereas only 40 to 50% of samples from surface waters contained detectable silver. Only one source, brine from a desert well, contained more than the permissible level of 50 μg/liter.

In a 1971 paper, Taylor cited a national water supply survey which found no silver levels over 30 μg/liter, and another study in which an average level of only 8 μg/liter silver was found in a wide range of waters. Since conventional treatment does not notably reduce trace elements levels, continued monitoring was felt to be necessary.

In a report to the Office of Water Research and Technology, Klusman and Edwards (1976) discussed toxic heavy metals in ground water of the Colorado front range. Although silver takes many forms in nature, such as argentite, pyrargite, and proustite, the most stable form in low sulfide environments was considered to be metallic silver. Oxidizing conditions may allow the formation of some Ag^+, but normally the levels will remain low. A sampling of water near the foothills showed an average of 0.58 μg/liter of silver, with 95% confidence limits of 0.13 to 2.64 μg/liter, well below the drinking water standard of 50 μg/liter. The probability of exceeding the standard both inside and outside the mineral belt was so low, it was assumed to be virtually nil.

9.8 CONCLUSIONS

Several major points have been demonstrated by the studies discussed in this chapter. It appears that most silver entering an aquatic system is complexed by organic matter in sediment and bottom muds and, in effect, is removed from the system. The pH of an aquatic system has been shown to influence the rate of silver adsorption, and can be helpful as a means of predicting the adsorption potential of a water. Sediments from a wide range of aquatic systems contain much higher amounts of silver than do their associated waters.

The silver content of both fresh and marine waters varies greatly, ranging from below detection limits to nearly 50 μg/liter. Only infrequently has a natural water been found to contain more than the drinking water standard limit of 50 μg/liter. In those waters having higher silver levels, neither plant nor animal life have been adversely affected, probably due to most silver being present in less reactive or less available forms. Numerous plants and primary consumer organisms in aquatic environments tend to accumulate silver, and many other metals, to levels higher than those of associated waters, and apparently silver uptake is not strongly regulated in some aquatic organisms. As some aquatic organisms can accommodate or become tolerant to the presence of metals, it will be difficult to predict long-term effects from repeated impositions of these materials on the structure and functioning of aquatic organism communities. There is little evidence, except for fish and aquatic invertebrates, that silver accumulates in higher trophic levels of ecosystems, even when secondary consumers and predators feed on primary consumers which have accumulated silver above levels present in the abiotic environment.

For many years, copper salts have been added to standing bodies of water in an attempt to control algal growth. Some consideration has also been given to the use of copper-silver mixtures, which might be more effective for this purpose. As with silver, copper is not considered a cumulative systemic poison.

10

Evaluation of the Potential Impacts of Silver and/or Silver Iodide on Rainbow Trout in Laboratory and High Mountain Lake Environments

P. H. Davies and J. P. Goettl, Jr.

10.1 INTRODUCTION

Davies et. al. (1978) reported that the toxicity of silver to rainbow trout (*Salmo gairdneri*) in soft water (hardness of 26 mg/liter as $CaCO_3$) was between 0.09 μg/liter (no mortality) and 0.17 μg/liter (17.2% mortality) in a long-term toxicity test of 18 months. A mean 96-hour LC_{50} of 6.5 μg/liter was obtained for silver in water of similar quality. Because of the extreme toxicity of silver nitrate to rainbow trout, and the current interest in the use of silver iodide as a weather modification agent, long-term tests were conducted to assess the possible toxicity of silver iodide. Silver iodide has a solubility constant of 1×10^{-16} M at 25°C, corrected to zero ionic strength (Sillen and Martell, 1964). This is equivalent to 1.08 μg Ag^+/liter, which is considerably higher than the toxic level for rainbow trout. At 15°C the solubility of silver iodide is 1.74×10^{-17} M (O.45 μg Ag^+/liter), still greatly exceeding the effect level.

Because of the insolubility of silver iodide in water, preliminary experiments used sodium iodide as a complexing agent. Ultimately, a method was devised by which the toxicity of silver iodide could be investigated within the limits of its solubility constant without using a complexing agent.

10.2 SILVER ANALYSIS

Initially, silver concentrations were measured by flame atomic adsorption spectrophotometric (AAS) analysis of ammonium pyrrolidine dithiocarbamate-methyl isobutyl ketone (APDC-MIBK) extracted samples (Fishman and Midgett, 1968) as modified by Davies (1975). Two hundred ml water samples were placed in 250 ml volumetric flasks, and following pH adjustment, the samples were chelated with 2.5 ml of 3% APDC and extracted into 10 ml of MIBK. Standards and blanks over an appropriate range were similarly extracted. This extraction method gave an analytical sensitivity of approximately 0.05 μg Ag/liter.

Flameless AAS analyses for silver were performed with a Varian Model 63, carbon rod atomizer using the "rifled" graphite tube. A total of 80 μl of each sample was injected into the graphite tube in increments of 40 μl, each of which was dried prior to ashing and atomization. This procedure allowed a silver detection limit of approximately 0.02 μg/liter.

10.3 SODIUM IODIDE AND POTASSIUM IODIDE ACUTE BIOASSAYS

Static 96-hour acute bioassays were conducted to determine the relative toxicities of sodium and potassium iodide to assess which salt would be the least toxic complexing agent for silver iodide. Experimental concentrations of each salt were 8,000, 4,000, 2,000, 1,000, 500, and 0 (control) mg/liter as NaI or KI. Ten fish, approximately 50 mm long, were exposed to each experimental concentration in 2 liter jars which were placed in a water bath at 14°C. Mortalities were monitored daily, and 96-hour LC_{50} concentrations were determined by log-probit analysis (Litchfield and Wilcoxon, 1949). The 96-hour LC_{50}'s for rainbow trout exposed to sodium and potassium iodide in soft water (30 mg/liter hardness) were 5,350 mg/liter and 2,190 mg/liter, respectively. Sodium iodide was selected as the complexing agent for use with silver iodide, since it was the least toxic.

10.4 SILVER IODIDE-SODIUM IODIDE LONG-TERM TOXICITY TEST

An experiment was initiated to test the long-term toxicity of silver iodide to rainbow trout. The silver iodide was made soluble by complexing with sodium iodide. A stock solution of 12 mg/liter silver iodide in 0.125 M sodium iodide was made by dissolving 192 mg silver iodide in 1 liter of 2 M sodium iodide which was diluted to 16 liters with demineralized water. Using a 50% proportional diluter (Mount and Brungs, 1967), nominal silver concentrations were 8.0, 4.0, 2.0, 1.0, 0.5, and 0.0 μg/liter, and nominal sodium iodide concentrations were 12.6, 6.3, 3.15, 1.6, 0.8, and 0.0 mg/liter. With sodium iodide used as a complexing agent, it was necessary to construct a diluter which would provide a sodium iodide control containing the same amount of sodium iodide as delivered with the highest silver concentration, 12.6 mg NaI/liter. The diluters delivered approximately 2 liters of test solution to each aquarium every 3 minutes, resulting in a 95% replacement time of the test solutions in 20 hours (Sprague, 1969).

Approximately 300 eyed rainbow trout eggs were used per aquarium. The eggs were placed in plastic hatching trays positioned under the inlet of the 265 liter fiberglass tanks with acrylic plastic covers, and the trays were monitored daily for dead eggs or fry. The experiment was terminated after six weeks of exposure.

A high mortality rate occurred with all silver concentrations and in the NaI control (Table 10-1). Eggs began to hatch about 10 days after initiating the test, but newly hatched sac fry usually died within a day or two of hatching at the higher exposure levels. Egg losses prior to hatching were approximately 10% in all tanks.

Sac fry in the dilution water control tank developed normally and had only 2% deaths during the six week exposure period. With 100% mortality in the sodium iodide control, it became evident that iodide ion contributed significantly to the deaths that occurred at all silver exposures. The highest concentration of sodium iodide added about 2.6 mg/liter of sodium to the ambient level in water, giving a total concentration of about 6.5 mg Na/liter which is well within the range of environmental levels. Therefore, the toxicity of iodide ion, which is present in excess of that needed to dissolve silver iodide, was significant. Water quality characteristics for alkalinity, hardness, pH, dissolved oxygen, temperature, and conductivity are given in Table 10-2.

TABLE 10-1. *AgI-NaI and NaI mortality data after hatching.*

	AgI-NaI chronic			NaI chronic	
Fish Aquaria	I^- (mg/liter)	Ag (μg/liter)	% Mortality	I^- (mg/liter)	% Mortality
#1	14.2	3.15	100	14.2	93.5
#2	6.8	1.98	95.2	6.8	79.4
#3	3.5	0.71	96.2	3.5	86.7
#4	1.5	0.37	71.6	1.5	72.6
#5	0.7	0.22	79.1	0.7	66.9
Control	0.0	0.00	2.2	0.0	1.9
NaI Control	14.2	0.00	100		

10.5 SODIUM IODIDE LONG-TERM TOXICITY TEST

This experiment was initiated at the same sodium iodide concentration used previously to assess the toxicity of iodide in the absence of silver, using nominal sodium iodide concentrations of 12.6, 6.3, 3.15, 1.6, 0.8, and 0.0 mg/liter. The experiment was initiated with 330 eyed rainbow trout eggs placed in hatching trays and was terminated after eight weeks. The iodide concentration was calculated from sodium concentration determined by AAS, which were corrected for background sodium levels found in the control or dilution water.

NaI exposure in the absence of silver also showed a high mortality rate, similar to that reported in the AgI-NaI experiment (Table 10-2). Eggs began to hatch about a week later than eggs in the preceding experiment, and the newly hatched sac fry again began to die a day or two following emergence in the higher exposure levels. Sac fry in the control tank developed normally and had about a 2% mortality rate. Egg losses were similar (6%) in all tanks. The addition of iodide (as NaI) apparently masked any possible toxic effects that might be attributed to silver iodide, making it necessary to work without a complexing agent.

10.6 PRELIMINARY SILVER IODIDE TOXICITY TEST

A preliminary experiment was set up to test the toxicity of silver iodide to rainbow trout within the limits of its normal solubility. A sealed PVC cylinder (25 cm inside diameter and 30 cm long) was constructed to permit water flow through dacron polyester filler material containing 700 g of dispersed silver iodide into a 35 liter aquarium. A 12 mm mesh PVC support grid at the top and bottom of the cylinder held the filler material in place and prevented blockage of the inlet and outlet. The flow rate to the aquarium was approximately 1 liter per minute, using water filtered through activated carbon. A control aquarium was set in a similar manner without silver iodide in the filler material. The experiment was initiated with 100 recently hatched rainbow trout sac fry in each aquarium.

This preliminary experiment demonstrated that silver iodide, within the limits of its own solubility, could be toxic to rainbow trout. Ninety-four

TABLE 10-2. *Water quality data for the AgI-NaI and NaI experiments.*

Item	Alkalinity mg/liter	Hardness mg/liter	pH	D.O. mg/liter	Temp. °C	Conductivity μmhos/cm
Mean	30	32	6.98-7.16	11.0	7.0	150

percent of the exposed fish died during six weeks of exposure, but no mortalities occurred in the control. The silver level of water passing through the supply cylinder was 0.88 μg/liter (range 0.5 to 1.1 μg/liter) versus less than 0.05 μg/liter in the control water.

In addition to mortality effects, very distinct retardation of development and growth was observed. After a week of exposure, the control sac fry had adsorbed the yold sac and were swim-up fry. However, complete swim-up did not occur with the fish exposed to silver until three weeks later. After six weeks the experiment was terminated, and at this time the control fish were considerably larger than exposed fish, 36 mm versus 24 mm, respectively. Chemical analyses for alkalinity, hardness, pH, dissolved oxygen, temperature, and conductivity were performed weekly (Table 10-3).

10.7 SILVER IODIDE LONG-TERM TOXICITY TEST, INITIATED WITH EYED EGGS

Following termination of the preliminary experiment, the silver iodide cylinder was rejuvenated by the addition of 250 g of silver iodide. Valves were used to adjust flows to water and chemical cells of the proportional diluter (Mount and Brungs, 1967) to give 50% dilutions for each of five concentrations and a control. The diluter delivered 2 liters of test solution to each aquarium every 3 minutes, resulting in a 95% replacement time of 20 hours (Sprague, 1969).

The experiment was initiated with approximately 340 eyed eggs per hatching tank. Following adsorption of the yolk-sac and swim-up, the fry were transferred to 265 liter tanks. To determine possible growth effects induced by silver iodide, the total weight of fish per tank was determined monthly. From these data, feeding rates were adjusted daily based on a monthly growth rate projection (Horak, 1969). Fish were randomly sacrificed at appropriate intervals to prevent overcrowding and to provide fish length information. Differences in fish growth after exposure to the various silver iodide concentrations were assessed using analyses of variance.

TABLE 10-3. *Water quality data for the preliminary experiment with silver iodide.*

Item	Alkalinity mg/liter	Hardness mg/liter	pH	D.O. mg/liter	Temp. °C	Conductivity μmhos/cm
Mean	15.2	12.7		8.5	15.9	99
Range	14-18	12-14	6.5-6.8	7.8-9.2	14-17	90-110

The "no effect" concentration for silver iodide was found to be between 0.03 and 0.06 μg/liter after 13 months of exposure (Table 10-4). In comparison, the "no effect" concentration for rainbow trout exposed to silver ion from silver nitrate lies between 0.09 and 0.17 μg/liter (Davies et al., 1978). Mortalities through swim-up of fry were essentially the same in all concentrations; however, following the embryonic stage of development, increased mortality occurred with increased concentration. The eggs began to hatch almost immediately and had completed hatching within eight days. At termination of the experiment there was no significant difference (P = 0.05) in growth between fish exposed to AgI and control fish (Table 10-5). Initial growth differences resulting from overcrowding in the control tank were not evident at the end of the experiment.

Chemical analyses on the experimental water for alkalinity, hardness, pH, dissolved oxygen, temperature, and conductivity were performed weekly during the 13 month exposure period (Table 10-6). There was no statistical difference (P = 0.05) between tanks for these analysis values.

10.8 SILVER IODIDE LONG-TERM TOXICITY TEST, INITIATED WITH GREEN EGGS

This experiment was conducted to determine the effects of silver iodide on the embryonic development, hatching and growth of rainbow trout. Eggs spawned from 10 female Hartman rainbow trout brood fish were fertilized with sperm from 10 males. Approximately 325 fertilized eggs were placed immediately into each of five silver concentrations and a control, consisting of 0.40, 0.18, 0.09, 0.04, 0.02, and 0.00 μg/liter. Growth data were collected at eight months and upon termination of the experiment after 10 months.

TABLE 10-4. *Mean silver concentrations and ranges, and mortality results (based on 310 rainbow trout per concentration) after 13 months of exposure.*

	Silver concentration (μg/liter)					
Item	0.41	0.21	0.11	0.06	0.03	0.00
Conc. range	.27-.85	.10-.42	.05-.22	.03-.10	.01-.05	.00-.00
% mortality (pre swim-up)	7.9	7.4	4.7	5.9	8.5	[a]
% mortality (post swim-up)	38.1	24.5	18.7	18.1	3.2	[a]

[a] Mortalities not given because of initial overcrowding of control tank.

TABLE 10-5. *Mean length of rainbow trout after 13 months of exposure to silver iodide (based on 40 fish/concentration).*

	Silver concentration (μg/liter)					
Item	0.41	0.21	0.11	0.06	0.03	0.00
Mean length (mm)	202.2	195.2	198.2	196.6	198.5	195.0
Std. deviation	16.3	18.0	17.2	16.6	14.0	17.2
Std. error of mean	2.6	2.8	2.7	2.6	2.2	2.7

Based on mortality data using 250 fish per concentration, the "no effect" concentration for rainbow trout exposed to silver iodide was between 0.18 and 0.40 μg/liter (Table 10-7). This "no effect" concentration is considerably higher than the "no effect" concentration for silver iodide reported for the eyed egg experiment, which was between 0.03 and 0.06 μg/liter. The higher "no effect" concentration of the present experiment might be explained as a result of the fish becoming acclimated to silver during the three months of embryonic exposure prior to swim-up. Hatching was completed in seven weeks (versus eight days in the eyed egg experiment), and swim-up was completed in 12 weeks (versus 17 days of exposure when initiated as eyed eggs). Similar accounts of embryonic acclimation to heavy metals have been reported for zinc (Sinley et al., 1974) and cadmium (Davies, 1976). Differences in genetic background for the various egg lots used may also account for different "no effect" ranges and mortalities observed at similar silver concentrations.

Embryonic exposure to silver iodide for 12 weeks failed to demonstrate effects of premature hatching found in an experiment where silver nitrate was used (Davies et al., 1978), or retarded sac fry growth as observed in the preliminary silver iodide experiment. The silver iodide experiment initiated with eyed eggs also failed to show these embryonic effects.

No significant differences in mortality occurred with eggs or sac fry exposed to silver iodide as compared with the control (Table 10-8). There also was no significant difference in growth at a 95% confidence level

TABLE 10-6. *Water quality characteristics during 13 months of exposure of rainbow trout to silver idodide.*

Item	Alkalinity mg/liter	Hardness mg/liter	pH	D.O. mg/liter	Temp. °C	Conductivity μmhos/cm
Mean	25.8	28.0	6.87	7.2	12.2	122
Range	14-36	20-34	6.5-7.3	6-10	6-17	85-160

TABLE 10-7. *Silver concentrations and mortality results (based on 250 rainbow trout per concentration) after 10 months of exposure.*

	Silver concentration (μg/liter)					
Item	0.40	0.18	0.09	0.04	0.02	0.00
Conc. range	.19-.52	.10-.26	.04-.13	.02-.06	.01-.03	.00-.00
% Mortality	26.8	2.0	3.2	2.0	3.6	2.4

between fish exposed to silver iodide and the control at 8 and 10 months (Table 10-9). The growth effect observed in the preliminary silver iodide experiment was probably a result of the higher silver concentration, 0.88 μg/liter, versus a high of 0.40 μg/liter in the experiment initiated with green eggs. Chemical analyses for alkalinity, hardness, pH, dissolved oxygen, temperature, and conductivity were performed weekly (Table 10-10).

10.9 EFFECTS OF SILVER ON A HIGH COLORADO ALPINE LAKE

As noted in the discussion of the laboratory studies of silver toxicity to fish, silver in the ionic form is clearly toxic to fish, whereas toxicity as silver iodide is more difficult to establish. In an effort to determine the possible effects of silver iodide on fish populations as related to weather modification activities, Buckeye Lake, a small alpine tarn of 9.5 m maximum depth, was studied for two years by Freeman (1975).

This lake is located in the southern Gore Range, Lake County, Colorado at an elevation of 3,627 m. Its watershed, which lies in the cloud-seeding target area of Colorado State University's Atmospheric Science Department, is composed of the western side of a large falling cirque on the southeast slope of Buckeye Peak. It is believed that Buckeye Lake has received representative doses of silver from seeding operations, since cloud seeding operations from 1964 to 1974 have released 19.738 kg of silver iodide over this area. If the Buckeye Lake watershed received at least its

TABLE 10-8. *Mortality effects of silver iodide on eggs and sac fry (based on approximately 325 rainbow trout eggs) during 12 weeks of exposure prior to swim-up of fry.*

	Silver concentration (μg/liter)					
Item	0.40	0.18	0.09	0.04	0.02	0.00
% egg mortality	1.7	2.1	5.8	2.6	7.5	7.7
% sac fry mortality	2.4	1.3	1.9	0.3	1.3	1.4

TABLE 10-9. *Mean lengths of rainbow trout after 8 months and 10 months of exposure to silver iodide (based on approximately 100 fish per concentration).*

	Silver concentration (μg/liter)					
Item	0.40	0.18	0.09	0.04	0.02	0.00
			8 months			
Mean length (mm)	90.0	88.4	88.6	90.6	91.2	88.6
Std. deviation	17.2	15.3	14.4	15.8	16.3	16.6
Std. error of mean	2.1	1.4	1.3	1.6	1.4	1.6
			10 months			
Mean length (mm)	131.3	128.4	129.2	124.2	132.1	124.0
Std. deviation	20.0	21.6	20.9	22.0	19.5	23.1
Std. error of mean	2.0	2.2	2.1	2.2	2.0	2.3

proportional share of silver from cloud seeding, and the estimations of Warburton (1971) are correct, then the Buckeye Lake watershed received between 0.43 and 1.80 g of silver in that 10 year period.

Preliminary measurements in the outlet indicated that a greater amount of silver was leaving Buckeye Lake than could be accounted for from weather modification activity. Bruland et al. (1974) indicated that the dust burden in the northern hemisphere from natural and man-caused activities was 2.0 μg Ag/m^3. This could cause an annual deposition of 0.4 μg silver per cm^2 from atmospheric scavenging. Freeman (1975) found the range of silver concentrations in the snow near Buckeye Lake to be from 0.02 to 0.44 μg/liter of water. The mean annual concentration, as water, was 0.42 μg/liter in 1973 and 0.37 μg/liter in 1974.

Another source of silver to Buckeye Lake may be from the soil in the watershed. Soils usually contain 0.25 to 0.50 mg/kg (Boyle, 1968). In comparison, the mean concentrations of silver in the lake water for 1973 and 1974 were 0.27 and 0.61 μg/liter, respectively. Silver concentrations with respect to water depth were irregular, and the degree of thermal stratification had no effect on silver concentration.

Seasonal trends in silver concentrations were observed in both years of the study. Peaks of silver concentration occurred at the end of the terrestrial plant growing season and following the spring thaw. It was postulated

TABLE 10-10. *Water quality characteristics during 10 months of exposure of rainbow trout to silver iodide.*

Item	Alkalinity mg/liter	Hardness mg/liter	pH	D.O. mg/liter	Temp. C	Conductivity μmhos/cm
Mean	25	29		7.2	13	144
Range	16-34	24-34	6.8-7.5	5.8-9.6	7-17	130-160

that at the end of the growth season large-scale die-offs and chemical reversals in the soil released higher silver concentrations to surface and groundwaters. Die-offs in the lake likewise commit greater amounts of silver to the sediments. For a short time during the fall, silver moves freely through the system without major biological involvement, until decreasing temperatures freeze the terrestrial water supply and seal the lake with as much as a meter of ice. Under the winter ice, sediment organic and silver levels remain high as a result of the accumulated biological fallout from the preceding fall. The silver levels in the water decreased, possibly via redox precipitation with iron, until the spring reoxygenation. With the onset of spring, surface and groundwater flows showed approximately the same silver levels as during the preceding fall. After spring thaw, biological activity both on the land and in the lake reinitiates the cycle.

The inlet stream silver concentrations illustrate these two peak periods even more dramatically than in the lake. The mean sediment silver values in 1973 were 0.49 mg/kg and in 1974, 0.28 mg/kg, further indicating a net loss of silver from the lake in 1974. In these sediments it was found that the silver concentrations was two to three times greater in clay-silt portions than in the sand and pebble portion. Silver concentrations at the sediment-water interface were more closely related to the concentration in lake water than to the sediment silver concentration.

Twenty cutthroat trout (*Salmo clarki*) were collected during the study period. These fish were analyzed for silver content by atomic absorption of wet pressure acid digested tissue (Adrian, 1971), and the majority of the silver was concentrated in the bones and liver, with decreased concentrations found with increasing age (Table 10-11). These fish fed primarily on chironomid pupae, the silver concentration of which generally followed that of the sediments in which they lived. No adverse effect was observed in a long-term laboratory feeding experiment using cutthroat trout fed the Colorado Standard Diet with silver iodide added to approximate the silver concentration observed in chironomid pupae.

In conclusion, Freeman (1975) described the ecological kinetics of silver in the Buckeye Lake ecosystem. He was unable to demonstrate toxic effects of silver on any of the components of this aquatic ecosystem.

TABLE 10-11. *Trends of silver concentration (mg/kg) in selected cutthroat trout tissues.*

	Scale age of fish		
Tissue	1+	2+	3+
Bone	4.42	2.78	2.56
Liver	2.32	1.29	1.39
Muscle	0.11	0.28	0.37
Skin	0.23	0.35	0.39

10.10 DISCUSSION

10.10.1 Toxicological Concerns

A "no effect" concentration of silver ion (from silver nitrate) of between 0.09 and 0.17 μg Ag/liter establishes silver, in the ionic form, as one of the most toxic heavy metals, and demonstrates the need for monitoring and enforcement agencies to analyze and report total silver concentrations in natural waters at a level of 0.10 μg/liter or less.

Coleman and Cearley (1974), in studies on the toxicity and accumulation of silver using largemouth bass and bluegills, reported a lethal effect between 7 and 70 μg/liter for largemouth bass, and no lethal effect for bluegills at 70 μg/liter during six months of exposure. The tolerance (no effect) levels for silver reported by these workers are misleading and do not reflect the potentially greater sensitivity of these species to more toxic forms of silver. The dilution water used in their long-term static experiments was relatively hard (180 mg/liter hardness) and contained an atypically high chloride concentration of 193 mg/liter. With chloride concentrations of this magnitude, most of the silver (added as $AgNO_3$) would be complexed into silver forms and species (as noted in chapters 3 and 4) which are, for the most part, not toxic.

Although silver iodide, when used in laboratory studies, can exhibit a toxic effect on rainbow trout similar to that found with silver nitrate, we do not think that the use of silver iodide poses a major threat to the aquatic environment. The long-term silver iodide toxicity tests used approximately 1,200 g of silver iodide in the polyester fiber filling the supply cylinder. Water flowed through the cylinder to the chemical cells on the diluter providing only that amount of silver capable of dissolving in the water. Theoretically, only about 0.45 μg/liter is soluble at 15°C.

It would be nearly impossible for silver iodide to dissociate sufficiently or rapidly enough to provide toxic levels of silver ion under the conditions of a flowing stream environment, even if sufficient amounts of silver iodide were released. Freeman (1975) estimated that only 0.43 to 1.80 g of silver was distributed over the Buckeye Lake watershed in a 10 year period of silver iodide use, even though the lake is located near the center of a weather modification target area.

10.10.2 Analytical Concerns

Heavy metals are frequently present in natural waters at concentrations lower than those measurable by direct determination by flame AAS. Frequently such metal concentrations can be toxic to the aquatic environment. Therefore, it is essential that low level analyses be performed if protection of the aquatic environment is to be achieved.

Organic extraction techniques provide an advantageous method for analyzing low concentrations of heavy metals using flame AAS. For some metals, sensitivities achieved by extraction methods are enhanced up to 200 times. Another advantage to the use of organic extraction is that sample treatment to remove interferences is not necessary, since alkali (Na and K) and alkaline earth (Ca and Mg) metals generally will not be chelated. The disadvantages of organic extraction procedures are that they are time-consuming, and contamination of extraction flasks by the adsorption of heavy metals can pose a serious problem

Direct flameless AAS using a carbon rod atomizer or graphite furnace also provides a method for analyzing very low concentrations of heavy metals in natural waters. A particular advantage of this method over the extraction method described above is that repeated sample analyses can be made immediately without re-extracting. In addition, contamination problems associated with the extraction technique are avoided. A disadvantage of the carbon rod atomizer technique is that the analyst must be thoroughly familiar with the instrumentation and technique, if reliable and reproducible results are to be obtained. Analyses generally require the use of background correction if various interferences are to be avoided. The operation of the carbon rod atomizer is also time-consuming compared to direct flame analyses.

Present research indicates that dissolved fractions (i.e., ionic and/or labile species) of heavy metals in natural waters constitute the toxic forms, and that the chelated and precipitated forms of heavy metals are not generally toxic while existing as complexed species (Davies, 1976; Davies et al., 1976; Andrew, 1976). The eventual goal for the aquatic toxicologist, aided by technological advances in analytical instrumentation, is to identify and evaluate the toxicity of particular heavy metal species in natural waters. Differential pulse polarography, differential pulse anodic stripping voltometry, ion selective electrodes, and other methods hold some promise for identifying and analyzing potentially toxic forms. Considerable research on methods and toxicological evaluation will be required before such analytical methods become applicable for use in defining and setting water quality standards and for monitoring and enforcement.

10.11 CONCLUSIONS

Long-term toxicity tests were used to assess the effects of silver nitrate and silver iodide on rainbow trout. An 18 month experiment gave a "no effect" concentration for silver (as silver nitrate) between 0.09 μg/liter (no mortality due to silver) and 0.17 μg/liter (17.2% mortality). In a 13 month experiment, using silver iodide within the limits of its own solubility (no chelating agent added) and initiated with eyed rainbow trout eggs, the "no effect" concentration was found to be between 0.03 and 0.06 μg/liter. Another silver iodide experiment initiated with green rainbow trout eggs gave a "no effect" level between 0.18 and 0.40 μg/liter. The difference between these two "no effect" ranges for silver iodide is probably the result of embryonic acclimation to silver where the fish were exposed as green eggs, making the fish more resistant to silver.

Although silver iodide can show a toxic effect on rainbow trout similar to that found with silver nitrate under specific conditions in the laboratory, we do not think that the use of silver iodide poses a threat to the aquatic environment. Considering the insolubility of silver iodide, it would be nearly impossible for silver iodide to dissociate sufficiently or rapidly enough to achieve toxic levels of silver in aquatic environments as a result of weather modification.

Silver concentrations in biological components of Buckeye Lake, Colorado, varied markedly through the season in response to changes in silver concentrations in the environment and seasonal growing conditions. Surface and ground water deliveries of silver to the aquatic ecosystem were regulated by the influence of climatic conditions on the terrestrial substrate. Variations in the net gain or loss of silver by the lake system as a whole were attributed to the efficiency of water-seston-sediment depositions and sediment resuspensions. Silver was found to constitute no imminent ecological threat under present levels of involvement in the subalpine Buckeye Lake ecosystem, located in an area where weather modification activities took place for a 10 year period.

11

Sewage Treatment Processes

D. A. Klein

11.1 INTRODUCTION

Sewage treatment, as discussed by Eckenfelder (1964), involves maintaining microbial biomass under controlled conditions to assure rapid biological utilization of dissolved organics. Aerobic and anaerobic processes are usually combined, giving the concept of "total oxidation" suggested by LaRiviere (1977). By combining both of these treatment procedures, one can maximize initial transformation of dissolved organic matter to microbial biomass and then transform the accumulated sludge (excess biomass) into methane and CO_2. The final residual material, when dried, has an acceptable texture and odor and can be used as a fertilizer.

Generally, sewage treatment produces sludges which have high metal concentrations. Much concern has been expressed that cadmium, lead, and nickel might enter the food chain of humans and other consumers if such sludges are used to fertilize food crops. Recent studies from California suggest that sewage sludge may be a valuable future source of precious metals (Frutig, 1977). Sewage sludge from Palo Alto was found to contain 21 ounces of silver per ton, together with other precious metals, and contracts for "mining" this metal source had been completed.

Unfortunately, as in many other areas of environmental concern, silver may or may not be measured in sewage analyses (Weaver et al., 1974; Van Loon et al., 1973), making it difficult to fully assess the concentration of silver in such materials. Due to the relatively short retention times for waste materials under aerobic conditions (2-4 hours), more attention has been paid to the possible toxic effects of various metals in the anaerobic digestion process which produces methane. This greater concern is based on the slower growth rates of methane-producing organisms (Wolfe, 1971), and the necessity for having retention times in anaerobic sewage treatment of at least 12-15 days (Pine, 1971) to maintain a stable system. As noted by Kugelman and Chin (1971), low concentrations (<1 mg/liter) of all free heavy metal ions (except iron) are extremely toxic, but sulfide can act as a precipitant. They emphasize that antagonism, synergism, and complexing reactions must be considered in reporting data on toxicity and stimulation. Poon and Bhayani (1971) studied the effects of a series of freely dissociable metal salts on substrate utilization kinetics by *Geotrichum candidum* and a mixed culture of sewage bacteria. They found that silver ion at 1 mg/liter could cause changes in the slope and K_m for both types of organisms under aerobic conditions. Kinetic analyses suggested that the inhibition was a simple linear noncompetitive system. They also noted that older activated sludges (48-72 hours) were less sensitive to the presence of toxic metals than were younger sludges.

Typically, a particular metal or a mixture of metals can have either antagonistic or synergistic relationships is sewage treatment processes, as noted by Kugelman and McCarty (1965). As in other related articles, the toxic metals discussed include copper, nickel, zinc, and chromium; no mention is made of silver.

11.2 AEROBIC VERSUS ANAEROBIC PROCESSES

The controversy over whether metals, including silver, will have greater or lesser effects on microorganisms with or without oxygen has continued for many decades. Cooper and Jolly (1970) postulated that silver would have greater effects upon anaerobic than aerobic microbial processes. Unfortunately, at that time it was not possible to predict the magnitude or direction of such a postulated silver effect. The earlier literature also suggests that this would not be an easy area for drawing definitive conclusions.

Heavy metals act both as stimulants and as inhibitors of anaerobic microbial processes. Bruni (1942) noted that metallic silver stimulated carbon dioxide evolution by yeasts in a quiescent culture (assumed to be anaerobic), and heavy metals have been suggested as a means of stimulating methanogenesis, the rate-limiting step in anaerobic digestion (Novak and Ramesh, 1975; Speece and McCarty, 1964).

The literature from central European laboratories suggests that under aerobic conditions metallic silver exhibits antimicrobial activities which are inhibited without oxygen. Doerr (1921) noted that metallic silver became antimicrobial only after reaction with oxygen. Zimmermann (1952) confirmed that silver metal was effective against microbes predominantly in the presence of oxygen, while silver ion was effective with or without oxygen. The formation of various silver oxides was felt to be the cause of increased antimicrobial activity after exposure of metallic silver to oxygen.

Additional available information suggests that metals, including various halide complexes, may be able to stimulate anaerobic digestion when present at relatively low levels. Koyama et al. (1970) noted that low levels (50 milliequivalents) of mixed marine water metals could stimulate methanogenesis but levels higher than 100 meq caused some inhibition, and with 300 meq, acetic acid accumulated in the test system.

11.3 PHOTOGRAPHIC PROCESS EFFLUENTS

Using photoprocessing effluents which contain silver predominantly in the halide complex or thiosulfate forms, Bard et al. (1976) found that silver in the tightly bound thiosulfate form did not have harmful effects on secondary (biological) sewage treatment processes. Other silver forms tested were silver sulfide and silver bromide. Successful biological treatment of such wastes has been carried out with silver concentrations in the range of 1 to 5 mg/liter. Using photoprocessing influents containing relatively higher silver levels, these workers found that 70 to 80% of the silver in the input stream of a sewage treatment plant is removed by the primary and secondary treatment processes. Influent and effluent values were 0.28 and 0.07 mg/liter, respectively. With such effluents, the 96 hour LC_{50} (50% lethal concentration) to flathead minnows was greater than 250 mg/liter (Terhaar et al., 1972). The Genesee River near the entrance to Lake Ontario did not contain high silver levels in spite of photographic industry activity in the area over the past 70 years. In nonindustrial effluents, Mytelka et al. (1973) found silver in only 6 of 385 effluent samples, using a 0.5 mg/liter detection limit.

With increasing use of tertiary systems for sewage treatment, recovery of silver from photographic effluents can be expected to improve. Gorbenko-Germanov and Baskova (1973) observed that silver halides tend to form stable, finely dispersed hydrosols which settle poorly. By oxidizing the organic complexes through ozonation, they facilitated the chemical precipitation of residual silver.

Linstedt et al. (1971) noted that advanced waste water treatment using filtration, activated carbon, and ion exchange could substantially remove silver and other metals from process streams. Radioisotopic silver was used as a tracer, and activated carbon and ion exchange efficiently removed from 97 to 99% of these metals. The final effluent value for silver was approximately 4.6 μg/liter, markedly below the USPHS standard of 50 μg/liter maximum silver content allowed in drinking water. It is important to note, however, that after using these techniques the problem of disposing of silver in the ion exchange wash solutions still remains.

In a general study of metal removal by conventional sewage treatment processes, Zemansky (1974) found that the percentage removal was greater at higher influent metal concentrations, which might be expected considering chemical engineering principles. Although silver levels were not monitored in these studies, the efficiency of metal removal decreased with lower metal concentrations in the input stream. This phenomenon is especially relevant because low silver levels are anticipated in waters influenced by weather modification.

11.4 EFFECTS ON METHANOGENESIS

Klein and Giangiordano (1976) studied the possible effects of silver iodide derived from three seeding agent burn complexes, AgI-NH_4I, AgI-NaI, and AgI-TiO_2 (Corrin and Barnes, 1976) on methanogenesis in anaerobic sewage sludges and muds. The sludge samples were incubated for two and four week periods, within the retention time range for normal anaerobic digester treatment (McCarty, 1971). After two weeks, the presence of silver in the various forms tested did not lead to significant decreases in methane production, with the possible exception of the 500 mg/kg treatments, where some depression of methanogenesis was evident. At four weeks of incubation, a similar effect was observed only with 500 mg/kg treatments. Similar analyses, using muds to which glucose was added to stimulate methane production, showed no significant differences between treatments and controls. However, there was a trend toward stimulated methane production with AgI additions, using muds which had been incubated for 1.5 and 2.5 year periods.

TABLE 11-1. *Effect of silver burn complexes on methanogenesis in sewage sludge[a] at two and four week incubation. (Klein and Giangiordano, 1976).*

		2 weeks			4 weeks		
Silver form	Silver level mg/kg	X	SD	Significance	X	SD	Significance
AgI-NaI Complex	1	0.79	0.11	—	1.16	0.16	—
	10	0.77	0.08	—	1.02	0.15	—
	100	0.79	0.07	—	1.07	1.10	—
	500	0.41	0.05	*	0.62	0.08	*
AgI-NH_4I-derived AgI	1	0.74	0.11	—	1.02	0.15	—
	10	0.79	0.10	—	1.18	0.13	—
	100	0.61	0.09	—	1.14	0.07	—
	500	0.44	0.02	*	0.57	0.04	*
AgI-TiO_2 Complex	1	0.68	0.06	—	1.01	0.07	—
	10	0.69	0.07	—	0.97	0.08	—
	100	0.68	0.08	—	0.97	0.13	—
	500	0.73	0.05	—	0.89	0.09	—
Sludge Controls		0.63	0.06		0.96	0.10	

[a]Possible significant decrease

These studies suggest that relatively high silver concentrations may be needed to cause changes in methanogenesis rates. This may be due to the formation of less reactive silver sulfides under reducing conditions.

Castignetti (1977) evaluated the effects of silver in the ionic and silver iodide forms from burn complexes on methanogenesis, using enrichment cultures of methane-producing organisms under defined cultural conditions. With silver ion, stimulation of methanogenesis occurred at silver levels of 32 to 64 $\mu g/ml$. With silver iodide from seeding generator burn complexes ($AgI\text{-}NH_4I$, $AgI\text{-}NaI$, and $AgI\text{-}TiO_2$), no effects were observed with these materials present at 1.0 mg/10 ml of culture medium.

11.5 SLUDGE DISPOSAL

Because silver and other metals concentrate in sewage sludges during waste treatment, the further disposal of these materials can be of environmental concern, depending on the disposal method selected. The general area of sludge treatment, utilization, and disposal has been discussed by Dick (1975). Although these wastes can be utilized as fertilizers and nutritional amendments, the problems of heavy metal transfer into various food chains is a point of continuing debate.

The disposal of sludge in estuarine or marine environments adjacent to major metropolitan areas can result in the accumulation of heavy metals in surface muds. Halcrow et al. (1973) measured high trace metal concentrations in a 20 km^2 area in the immediate vicinity of a sewage sludge deposit. Silver was present at 3 $\mu g/g$ only in a restricted area, and in adjacent nontreated areas silver was generally present at 0.2 $\mu g/g$ levels. Studies from Norway have noted a general relationship between silver and copper levels in sediments in an industrial effluent-affected area (Lande, 1977). Copper and silver, in addition to having similar biological characteristics, have also been used in combination as a nucleating agent (Passarelli et al., 1973) and as an algicide (Young and Lisk, 1972).

11.6 CONCLUSIONS

Sewage treatment tends to concentrate heavy metals, increasing the possibility of effects on microorganisms essential to these processes. The long-standing controversy as to whether silver will have greater impacts on aerobic or anaerobic systems has led to much research in this area. Oxygen has been shown to be important in potentiating the antimicrobial capabilities of silver and silver halides.

Photographic processing effluents containing silver thiosulfate have not been shown to harm secondary sewage treatment processes, and such compounds can be largely removed from process streams.

Silver iodide from seeding agent burn complexes can inhibit methanogenesis, but only at extremely high (500 mg/kg) levels, and low levels of silver ion (32-64 μg/ml) can stimulate methanogenesis under defined cultural conditions.

Disposal of sludge probably requires further investigation, since its use as a fertilizer may result in returning silver to the food web.

As silver contributions to sludges due to weather modification activities are minor compared with much larger contributions from other commercial sources, detecting their impact will be difficult.

12

Effects on Humans*

D. A. Klein

12.1 INTRODUCTION

Since the early 1930s, concern about possible detrimental effects on humans from heavy metal (and specifically silver) exposure has led to a great deal of research. Historically, the scientific community has studied silver concentrations in biological materials, speculated about toxic effects of silver used in medicine and water purification, and occasionally observed argyria, a gray pigmentation of the hair and skin caused by deposition of silver in the tissues.

In 1940 Kehoe et al. analyzed the trace metal content of normal human tissues, blood, and urine. Silver was present in 10 to 20% of the samples assayed, and in quantities generally one or two orders of magnitude less than those of manganese, lead, tin, aluminum, and copper. Specifically, silver concentrations ranged from 0.002 to 0.005 mg/100 g tissue in the brain, liver, lung, and intestines. Only rib bone samples had levels as high as 0.01 mg/100 g tissue, and other structures such as kidney, heart, muscle, long bone, and stomach tissue contained no detectable silver. Urine and whole blood analyses did not show silver with the methods used. Food and fecal silver levels studied over a 24-hour period after ingestion, indicated almost complete elimination and scant absorption by alimentary tissues within this time. The authors noted that silver in the diet is largely due to food containers, utensils, and tableware containing silver.

Douglas (1968) discussed possible toxic effects of weather modification materials. Precipitation from storms seeded with silver iodide had silver concentrations ranging from 20 to 200 $\times$ 10^{-12} g/ml, equivalent to 0.02 to 0.2 μg/liter. These levels were considered to be too low for human metabolic system response to their presence. Douglas listed argyria as a toxic effect of continued exposure to silver compounds, but noted that there were no specific toxicity data for silver iodide. Regarding airborne

*Major contributions to the writing of this chapter were made by Gayle Robison.

silver iodide near a generator, he pointed out that the mass of silver iodide is 10^{-9} to 10^{-11} g/m^3, which is 10^{-5} of the mass concentration permissible for occupational exposure. Since it is unlikely that anyone would remain near a seeding agent generator for an extended period of time, the only potential hazard would be exposure to iodine vapor.

Douglas (1970) considered the possible effects on humans of silver iodide generator operation in greater detail than in his previous article. The iodine vapor level near a generator is approximately one-half the threshold level of 1.0 mg/m^3, so it was not considered a human health hazard. The carbon monoxide level of 150 μg/g can be tolerated for 1 hour with no effects whatsoever. Silver iodide and associated complexes are not likely to be absorbed by humans near a generator, due to their low concentrations and extremely low solubilities. Even if absorbed, Douglas felt that no symptoms would result. He concluded that any materials released to the atmosphere by a silver iodide generator may be ignored due to the small quantities involved and the low toxicity of the compounds released.

Standler and Vonnegut (1972) suggested that silver iodide in the air and in precipitation poses no danger to humans. Exposure to 2 to 50 grams of silver iodide resulted in argyria in patients studied, but this disease appears to be solely cosmetic. General toxicology references do not list silver iodide as poisonous, although Sax (1963) lists its dangerous properties as typical of a silver salt. Standler and Vonnegut emphasized that one could drink water produced by cloud seeding for a lifetime and not develop argyria. No ill effects (aside from rare cases of rash and aggravated allergy) were noted by these authors in those who operate silver iodide generators, according to their interviews with seeding generator operators. Persons handling the concentrated silver iodide solutions reported yellow staining of the skin, but no toxic effects.

Studies of argyria reveal potential effects of large-scale ingestion or exposure to silver in various forms. Blumberg and Carey (1934) reported a unique case of obscure, undetected argyria which was diagnosed after detection of high blood silver levels. The patient exhibited a peculiar gray pallor overall and gray-blue pigmentation in the gum margin around her teeth; both symptoms were apparently due to early argyria. In addition, she was anemic and asthenic, which the authors reported was due to silver intoxication. For a year the patient had taken 48 mg silver nitrate per day for alternate periods of 14 days. Blumberg and Carey postulated that light may play a role in production of the argyria pigment, which would explain the retarded appearance of discoloration in the patient, who was not exposed to strong light during the silver treatment.

Buckley (1963) discussed localized argyria in a study of discoloration confined to the fingers of a woman exposed to silver fixing solution for several months. It is essential to note that the discoloration appeared only where the skin had been in direct contact with the solution. Studies of skin specimens showed that the staining did not extend much below the epidermis, and was continuous. Buckley concluded that this compound was melanin, and speculated that silver ion diffused through the sweat glands to adjacent papillary bodies, where it became lodged and stimulated melanin production. The rarity of the finding, he explained, is due to the normally low (1 to 2 g/liter) concentrations of Ag involved, and the fact that operators usually wear gloves when working with fixer.

In further studies of localized argyria, Buckley et al. (1965) determined that the silver-containing particles involved were Ag_2S. Most of these particles were distributed along the dermoepidermal junction, and occurred intracellularly. The authors noted that the mechanism of increased melanin production was unknown.

Buckley and Terhaar (1973) suggested that the skin should be considered an excretory organ for this particular metal. Silver was found to move from the general body pool through the dermis and into the epidermis. In this process, silver was released from melanin-silver complexes as soluble ionic silver near the surface of the epidermis.

In the book *The Geochemical Environment in Relation to Health and Disease*, Hopps and Cannon (1972) compiled data on health problems and possible relationships between trace inorganic compound levels and health.

The *U.S. Registry of Toxic Effects of Chemical Substances* (Christensen and Luginbyhl, 1975) gives specific details on silver toxicity. The lowest recorded toxic concentration of silver is 1 mg/m^3 (inhaled), while that of colloidal silver is 700 μg/kg body weight. The U.S. occupational standard for both substances is 10 μg/m^3 (inhaled). Colloidal silver has a reported LD_{50} of 100 mg/kg (oral mucosa). Little information is given about silver nitrate toxicity, but the LD_{50} is said to be 50 mg/kg (oral mucosa), and the U.S. occupational standard is 15 μg/m^3.

Just and Szniolis (1936), in a frequently-quoted paper on the germicidal properties of silver in water, stated that compounds of heavy metals could be harmful to humans, based on immunological changes observed in test animals. However, they felt that less than 50 μg/liter of silver in waters would cause no observable effects.

Russian scientists, as noted by Jasionowski and Allen (1974), chose silver as the ideal long-term sanitizing agent for space ships and space stations. Up to 0.2 mg/g silver in drinking water has caused no detrimental effects on experimental animals. In addition, the year-long use of water containing high silver levels did not lead to medical problems or physiological changes in human subjects.

12.2 SILVER IN TISSUES

Recent studies have revealed small amounts of silver in the tissues and organs of healthy humans. Tipton and Cook (1963) found silver in approximately half of all samples tested. Of the 29 major tissues and organs examined, 24 contained a median of 0.1 μg Ag/g ash or less.

An environmental study of the concentration and distribution of stable elements in healthy human tissue (Hamilton et al., 1973) reported levels of silver ranging from 0.001 μg Ag/g wet weight in lymph nodes to 1.1 μg Ag/g wet weight in rib bones. Generally, blood samples contained only 0.01 ± 0.005 μg/g wet weight of blood while in the brain and other internal organs, similar or even lower values were noted (Table 12-1). However, a value of 0.3 μg/g wet weight was noted in work by Anspaugh which was presented in this table. Without dry weight conversions, it is difficult to draw conclusions from these data on possible bioconcentration.

Diem (1962) found that the blood concentration of silver on a dry weight basis could be considered to be equal to or only slightly higher than that for the crustal material, implying that biocentration was not significant.

Hamilton and Minski (1972) determined the trace inorganic composition of DNA in calf thymus and fish sperm, and of yeast RNA, to determine the actual sites of metal retention. As shown in Table 12-2, yeast RNA silver levels ranged from 0.02 to 0.1 μg/g dry weight, and calf thymus DNA levels were somewhat lower. Herring and salmon sperm DNA contained slightly more silver than the other sources. The authors emphasized the importance of this kind of study, noting that more information is

TABLE 12-1. *The concentration of silver in various portions of human materials (Hamilton et al., 1973).*

Component	Number of Samples	Silver Concentration μg/g dry weight
Brain—Whole Brain	10	0.004 ± 0.002
Frontal Lobe	2	0.003 ± 0.001
Basal Ganglia	2	0.004 ± 0.0002
Kidney—Entire	8	0.002 ± 0.0002
Cortex	8	0.001 ± 0.0002
Medulla	8	0.002 ± 0.0002
Lymph Nodes	6	0.001 ± 0.0002
Lung	11	0.002 ± 0.0001
Muscle	6	0.002 ± 0.0005
Testis	5	0.002 ± 0.0004
Ovary	6	0.002 ± 0.0005
Rib Bone[a]	22	1.1 ± 0.2

[a]Hard and soft water areas.

TABLE 12-2. *The chemical (inorganic) composition of DNA, RNA in $\mu g/g$ dry weight (Hamilton and Minski, 1972).*

Element	DNA (Mg salt) Salmon Sperm	DNA Herring Sperm	DNA (Na salt) Calf Thymus	DNA Calf Thymus	RNA Yeast	RNA Yeast
Ag	0.05	0.04	0.02	0.003	0.02	0.2
Cu	0.8	6.4	4.0	1.1	14.2	69.0

needed on tolerance to trace elements and on long-term effects of exposure.

In a discussion of the elementary composition of living matter, Bowen (1966) gives the average level of silver in mammals as 0.006 $\mu g/g$. The brain contained the highest level of silver in mammals (0.04 $\mu g/g$), while muscle averaged only 0.004 $\mu g/g$.

The diet of an average *Homo sapiens* was considered to contain 0.06 to 0.08 mg silver per day, in comparison with a toxic dose of 60 mg/day and a lethal dose of 1,300 mg/day.

12.3 ABSORPTION AND EXCRETION

Studies by Durbin (1960) and Furchner et al. (1968) indicate that as much as 10% of ingested silver is absorbed by the body, but that only approximately 4% is retained in the tissues. It becomes apparent then that large quantities of silver must be ingested in order for possibly toxic levels to accumulate in the tissues. Furthermore, Kehoe et al. (1940) showed that most silver will be excreted in the feces of persons ingesting less than 0.1 mg in one day. It is likely that any silver absorbed due to low-level exposure would be removed almost as soon as it is ingested with only traces accumulating in the tissues.

Kehoe et al. (1940) reported excretion of 65% of ingested silver within 24 hours after oral administration, but it was not until the next year that detailed studies of "minor" element absorption and excretion were made.

To obtain information on silver metabolism, Kent and McCance (1941) studied a patient suffering from argyria. They noted that the patient excreted very little silver, confirming other studies which indicate the permanence of argyria pigmentation. An investigation of average amounts and variations of trace elements in ordinary diets (Tipton et al., 1966) showed both positive and negative silver balances, but emphasized that silver, once fixed in the tissue, will be excreted slowly.

In the U.S. Public Health Service *Drinking Water Standards* (Anonymous, 1962), the need for setting standards for silver in drinking water is discussed. McKee and Wolf (1963) noted that although the total dosage of ingested silver required to produce argyria is unknown, the intravenous injection of 1 gram will cause this effect. Drinking 2 liters of water per day, with a drinking water standard limit of 0.05 mg/liter, 27 years would be required to ingest 1 gram of silver, most of which would be eliminated through the digestive tract. A review of 200 cases of argyria showed consistent delays in appearance of discoloration and a wide variation in predisposition to the disease. Apparently mild systemic effects are over-shadowed by external changes, but there is no evidence to substantiate this. Silver is considered to be excreted primarily by the liver, and is rapidly eliminated when given carrier-free. Despite the uncertainties involved in considering silver in water used for culinary and bathing purposes, it was stated that 10 μg/liter could be ingested for a lifetime before 1 gram silver would accumulate from 2 liters of water ingested per day. It was emphasized that water supplies containing >50 μg/liter should be rejected due to the possibility of long-term accumulation leading to delayed harmful effects.

In addition to silver ingestion via food and water, it has been shown that silver and presumably silver iodide may easily enter the body through the lungs. Phalen and Morrow (1973) demonstrated translocation and clearance of radioactive silver inhaled by dogs. They state, however, that the real hazards from exposure to nonradioactive silver compounds today are essentially negligible.

In a 1973 study of trace elements in particles emitted from coal-powered electricity-generating plants, Natusch et al. (1974) found that many elements such as Pb, Zn, Ni, Cd, and Cu were present at highest concentrations in the smaller (0.5 to 1.0 μm) particles, 50 to 80% of which are absorbed in the alveolar regions. Larger particles deposit in the upper regions of the respiratory tract where they can be removed by cilial action, and eventually reach the stomach where absorption efficiency is only 5 to 15%. Standler and Vonnegut (1972) assume that 0.1 μm AgI particles will be produced in the process of nucleant generation, so absorption of the components could reach 50%, but with levels of only 0.1 ng/m^3 (10^{-5} of maximum allowable) there appears to be no cause for concern, even for persons standing next to a generator for a prolonged period of time.

12.4 SILVER AND CHRONIC DISEASES

Since 1960, additional detailed studies on possible links between trace metal exposure and diseases in humans have been published. Schroeder (1960) discussed relationships between 20 trace metals and chronic diseases. Postulating that any element not found in plants is not essential for man, Schroeder stated that since Ag and Au are not usually found in plants they may be classified as environmental contaminants without known functions. In the section on specific metal displacements, he noted the likelihood that Cu^+, Ag^+, and Au^+ will compete. Silver presence was not considered to be linked to arteriosclerosis, cancer, collagen diseases, or allergies; and the author points out that urine silver levels in hypertensive persons are normal.

Fujita (1972) studied the carcinogenicity of an Ag-Pd-Au alloy imbedded subcutaneously. It was found that a 10 × 10 mm nonporous plate of the alloy induced tumors in 50 of the rats, whereas only 5% of the porous plates induced tumor growth. The author concluded the carcinogenicity was due to chronic stimulation, and not to the specific elements involved. To date, this is apparently the only link between silver and malignant tumors. It must be noted that AgI from cloud seeding will not normally form an alloy with Pd and Au. Should it form such an alloy, it is highly improbable that it could become subcutaneously imbedded for the average of 500 days required to induce tumor formation.

12.5 CONCLUSIONS

The effects on humans of AgI used in cloud seeding are minimal. Most human tissues contain silver, and judging from past difficulties in determining environmental silver level changes due to weather modification programs, an increase in tissue silver due to the use of seeding agents would not be detectable. Silver can accumulate in tissues due to environmental exposure although the levels are extremely low, with only about 4% of ingested silver actually being retained in the tissues. Argyria would probably never result from ingestion of the small quantities of silver introduced to the environment by weather modification activities.

Absorption of silver occurs primarily via ingestion or inhalation. Ingestion of trace amounts of silver normally present in the diet appears to be followed by excretion of slightly lower amounts, resulting in minimal accumulation. Water supplies may be a potential source of increased silver exposure but the legal limit of 0.05 mg/liter, if not exceeded, will restrict ingestion from this source. Inhalation of Ag is followed by a distribution of the element throughout the body; excretion follows, but some silver remains in the body. Detailed studies of trace elements in relation to chronic and malignant disease have not implicated silver as a significant concern. The only experimental link between silver and cancer emphasizes the improbability of silver causing harmful effects in humans. Argyria, a chronic discoloration of the skin, is caused by continued exposure to silver, but even those who are in continual contact with seeding generators have not experienced this problem.

13

Silver Effects: Other Terrestrial Organisms

D. A. Klein

13.1 INTRODUCTION

Although silver in low concentrations appears to have essentially no physiological effects on man, it is necessary to study possible silver effects on other terrestrial organisms which might have critical functions more sensitive to this metal. Matrone (1969) attempted to predict effects of elements on animal metabolism, based on their d electron relationships. Divalent silver was predicted to interfere with copper but not with zinc metabolism (Table 13-1). Davis (1972) also obtained results supporting the hypothesis that silver would have effects similar to copper. Silver at 100 μg/g reduced growth in copper-deficient chicks, but not in control chicks, and it was assumed that silver acted as a copper antagonist. In addition, silver tends to accentuate the effects of copper deficiency (Van Campen, 1966). Apparently silver in itself is not harmful, and only in combination with other stress factors might be detrimental. Even then, the effects appear to be reversible.

TABLE 13.1 *Chemical characterization of metals having characteristics similar to silver (Matrone, 1969).*

Ion	"d"	Configuration	C.N.*
CU^{+}	d^{10}	sp^3	4
Zn^{++}	d^{10}	sp^3	4
Cd^{++}	d^{10}	sp^2	4
Hg^{++}	d^{10}	sp	2
Cu^{++}	d^9	dsp^2	4
Ag^{++}	d^9	dsp^2	4

*Coordination number.

13.2 DISTRIBUTION IN OTHER HIGHER ORGANISMS

The silver levels in many animal-derived materials are available. The concentrations of silver in beef steak and beef heart RNA have been determined by Mitteldorf and Landon (1952). Silver was present in amounts ranging from 0.004 μg/g in bulk grade meat and 0.007 μg/g for choice grade to 0.024 μg/g for utility grade samples. Trace elements are known to react with nucleic acids and Wester (1972) determined the trace element levels in beef heart RNA. An element considered to have no known biological function, silver was found in levels of 0.12 to 0.45 μg/g RNA. The amounts in whole heart tissue ranged from 0.001 to 0.025 μg/g dry weight.

The absorption and excretion of silver have also been studied extensively in rats. In a major study, Scott and Hamilton (1950) administered radio-silver to rats intramuscularly, intravenously, and by stomach tube. Following administration of 10^{13} atoms (about 10^{-10} M) of silver, excretion by renal processes was thorough (Fig. 13-1). Less than 0.1% of silver administered intragastrically was absorbed. The distribution of silver after intravenous injection was the same as the distribution after intramuscular injection. With a dosage of 10^{18} silver atoms per rat, a concentration similar to that used in therapy, a different distribution was obtained. A 10 to 100-fold concentration increase of silver was observed in the liver, kidney, spleen, skin, bone, and muscle. Rats with anesthesia-damaged livers were unable to excrete an intramuscular dose of silver containing as few as 10^{13} atoms. However, the damage was transient and the animals recovered within a few days. This normal route of silver excretion was impaired by bile duct ligation in later experiments, resulting in silver accumulation in liver and body tissues. Some silver was excreted in the feces after ligation.

Silver localization in rat lysosomes was also noted by Brun and Brunk (1968), based on histochemical evidence. They felt that heavy metals are normal constituents of lysosomes "in a more generalized way" than had been previously assumed, although this was not discussed further.

In studying the effects of silver on the absorption and distribution of radio-copper in rats, Van Campen (1966) determined that silver decreased the relative proportion of ^{64}Cu in the blood and increased the proportion deposited in the liver (Table 13-2). Silver reversibly inactivated some copper-containing enzymes, bound sulfhydryl groups, and was concentrated in the liver. These properties could affect the normal pathways of copper metabolism, as noted in earlier studies.

Whanger and Weswig (1970) found an interesting relationship between copper and silver in that silver is the strongest copper antagonist, followed by Cd, Mo, Zn, and SO^{4-}, in a study where ceruloplasmin levels were measured.

Scott (1967) used labeled silver to study the possible movement of metals across the blood-brain barrier of rats. Rats were exposed to chronic

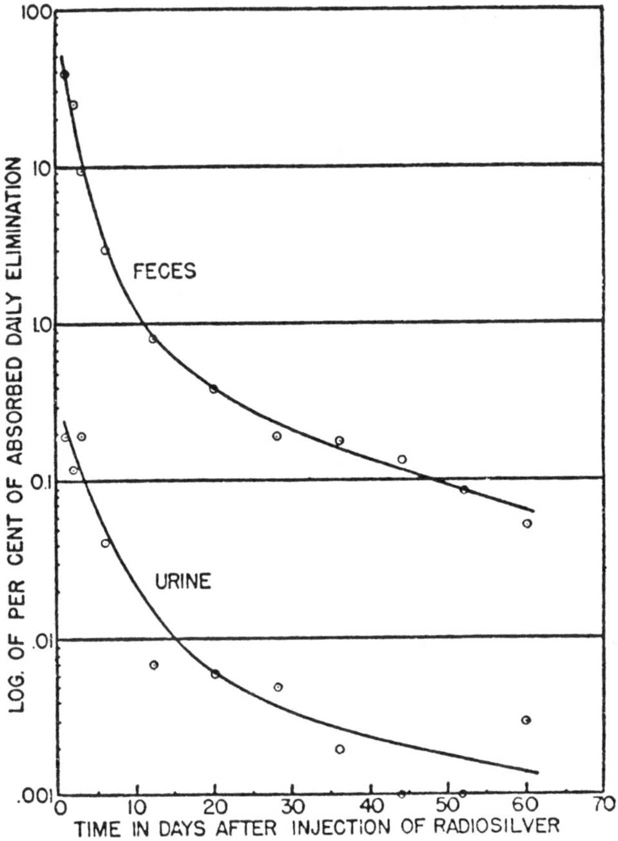

FIGURE 13-1. *Fecal and urinary elimination of radio-silver in rats (Scott and Hamilton, 1950).*

low-level radiation and silver nitrate solution was administered. The blood-borne silver contributed insignificantly to brain silver levels (Fig. 13-2), and silver was deposited in areas which did not influence the blood-brain barrier, primarily in the large superficial blood vessels. Silver was found in similar concentrations in radiation-treated and control animals.

Further studies on the uptake and absorption of silver support the work of Scott and Hamilton (1950). Furchner et al. (1968) examined the retention of ^{110}Ag administered orally, intravenously, and intraperitoneally to rats, mice, and other animals. Ninety percent or more of the oral doses were not absorbed, and most of the material (99.9%) was excreted in the feces within 15 days. Anghileri (1969) investigated the distribution of insoluble radio-silver iodide after intravenous injection using rats and mice. All of the halides were broken down rapidly, and the radio-silver

TABLE 13-2. *Effect of silver on the absorption of distribution of* ^{64}Cu (Van Campen, 1966).

Ag/Cu	% of dose in sampled tissues[1]	Relative distribution[2,3]				
		Blood	Heart	Kidneys	Liver	Spleen
		%	%	%	%	%
0	21.89[a]	24.47[a]	0.42[a]	17.09[a]	58.06[a]	0.42[a]
130	14.12[a]	18.07[ab]	0.33[a]	15.26[a]	65.85[ab]	0.48[a]
300	12.71[a]	16.96[ab]	0.42[a]	16.92[a]	65.36[ab]	0.34[a]
600	19.32[a]	11.54[b]	0.26[a]	15.91[a]	71.87[b]	0.47[a]

[1] Percentage of administered dose recovered in blood, heart, kidneys, liver and spleen.
[2] Each value is the mean of 5 observations.
[3] Values in the same column which are not followed by the same letter are significantly different ($P \leq 0.05$).

iodide was released as a nonionic form excreted via the bile. There appeared to be a preferential retention of silver and iodide in different organs, especially of iodide in the thyroid.

Mice and rats were used by Gammill et al. (1950) to determine the distribution of radioactive silver colloid in tissues, following infection by

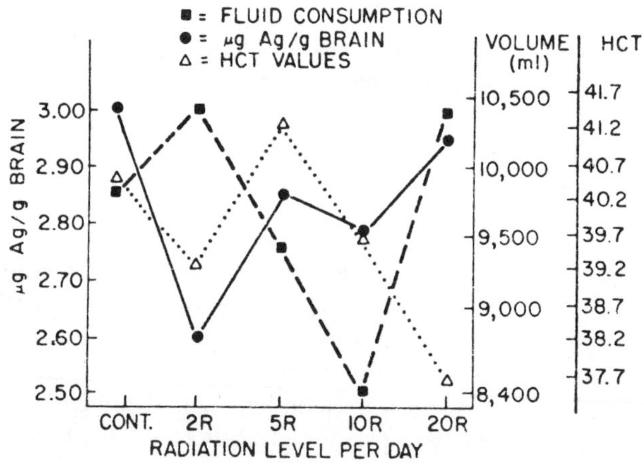

FIGURE 13-2. *Average volumes for rats exposed to 2, 5, 10, or 20 R per day and control animals: silver concentration per gram of brain tissue, hematocrit, and volume of silver nitrate solution consumed over 90 day test period (Scott, 1967).*

various routes. The mice were injected intraperitoneally and the rats were injected intravenously, intramuscularly, and intracardially. Intramuscular injection resulted in a high concentration of the silver at the injection site. Intravenous and intraperitoneal injections resulted in the deposition of silver in the reticuloendothelial system. A considerable amount of silver was eliminated through the intestinal tract.

13.3 PHYSIOLOGICAL EFFECTS ON OTHER HIGHER ORGANISMS

There is conflicting evidence as to whether Ag-Pd-Au alloys can cause tumors. Fujita (1972) noted that nonporous Ag-Pd-Au alloy plates embedded subcutaneously in rats induced tumors in 50% of the animals after 504 days. However, Habu (1968) implanted Ag-Pd-Au alloy pellets in the suboral mucosa of rabbits, and into the liver, testicles, and femoral muscle of rats, and after 90 days noted no effects related specifically to silver presence. Guyer and Mohs (1933) experimented with intratumoral injections of colloidal platinum and silver in rats. Rats whose tumors were injected with platinum showed tumor growth retardation and healing. As a control for the platinum experiment, a colloidal suspension of silver was also injected into tumors. Of the silver-treated animals, only one (2.5%) healed while five (16.6%) of the controls healed. They concluded that colloidal suspensions of silver injected intratumorally did not inhibit the growth of cancer, but may have stimulated it. Oppenheimer et al. (1956) embedded silver, steel, tantalum, tin, and vitallium foil into the abdominal walls of six sets of 25 rats each. After 300 days, 32% of rats with silver implantations developed fibrosarcomas, and silver produced more tumors than any of the other metals (Table 13-3).

Olcott (1950) fed rats various metal salts in their drinking water, another method of assessing possible effects of high level exposure. Two hundred and forty-five rats were given silver salts in a 1:100 concentration

TABLE 13-3. *Results of imbedding metals in rats (Oppenheimer et al., 1956).*

Material	Latent period (days)	No. malignant tumors	No. days since imbedding*	No. rats still alive
Silver	275-625	14 (32 per cent)	(Completed)	None
Steel	364----	5	596	4
Tantalum	714----	2	844	2
Tin		0	(Completed)	None
Vitallium	453----	5	725	2

*Based on 30 imbeddings with each metal, two in each of 25 rats.

instead of plain water, while 188 control rats were used. No arteriosclerosis or other vascular changes could be attributed to silver; however, hypertrophy of the left ventricle occurred in a statistically significant number of animals treated with silver over a long period of time. He postulated that this condition was caused by vascular hypertension due to silver deposition which thickened the basal membranes of the renal glomeruli.

Wahlberg (1965) investigated the percutaneous toxicity of metal compounds in guinea pigs. Of the compounds tested, $AgNO_3$ was found to be one of the least toxic. None of the experimental animals treated died from intraperitoneal injections or skin applications of the compound. However, the compound did lead to decreased weight gains.

Silver in combination with a vitamin E deficiency can apparently lead to liver necrosis and death in the rat (Diplock et al., 1967; Grasso et al., 1969), as it can in chickens. The latter study postulated that without vitamin E, silver could replace selenium as "selenide" in membrane iron proteins, which would limit their ability to function as electron carriers and disrupt the functioning of a variety of subcellular organelles, including the lysosomes. Silver acts mainly as a liver poison against which vitamin E, DPPD, and dietary protein, at high concentrations, are protective (Bunyan et al., 1968).

Mice, rats, and rabbits were subjected to silver in a number of experiments in order to determine what concentration of silver in drinking water might be toxic (Barkov and El'piner, 1968). In long-term experiments on rabbits, doses of 0.25 and 0.025 mg of silver per kilogram reduced immunological activity. The same doses in rats affected the animals' conditioned reflexes, while dosages of 0.0025 or 0.00025 mg/kg had no detrimental effects. The toxicity of silver is relative to body weight, with 0.0025 mg silver per kilogram of body weight being the maximum ineffective dose for most animals. These authors recommended that if silver ion is used to preserve water on sea-going vessels, the silver level should be kept at less than 0.05 mg/liter, which approximates the maximum ineffective dose. This incidentally is the U.S. Public Health Service limit for silver in drinking water.

Histological studies of silver injection into rats or mice have been carried out by many workers. Usually, with a sufficient level of free silver ion, it is possible to observe histological changes or effects. Typical experiments have been concerned with possible silver ion effects on the histochemistry of rat testes (Hoey, 1966), and of silver sulfadiazone chemotherapy and toxicology in DF-1 mice, where *Plasmodium* and *Pseudomonas* control was investigated by Wysor (1975a). No abnormal reactions or pathology were seen in mice receiving 1050 mg/kg of silver orally or subcutaneously each day for 30 days. However, when administered interperitoneally, acute systemic toxicity was observed.

13.4 EFFECTS ON RUMEN AND CAECUM FUNCTIONS

The major work in this area has been completed by Roy and Bailey (1974), in which silver ion, silver iodide, and silver iodide burn complexes were used. Silver iodide complexes at a concentration of 38 μg/g silver in dry alfalfa did not affect substrate digestion *in vitro*. With 100 μg/g of silver as silver ion, an inhibition of rumen digestion was noted, as might be expected. In addition, chronic intake (using goats) of 1.0 μg/g in feeds did not lead to silver accumulation, and the elimination half-time was 1.3 days. The presence of chlorides, calcium ion and organic matter, and short retention times in the rumen make it unlikely that low levels of silver iodide, if they might be present in animal feeds, would lead to inhibition of rumen functions. Martz et al. (1974) noted that silver used as a tracer in ruminant studies can become attached to plant cell wall particles of ingested feed materials. The particles absorbed over 67% of the added silver, which could also lead to decreased retention times and lower bioavailability in the rumen. Using enriched cultures of methanogens from the rumen in defined media, Castignetti (1977) obtained similar results using silver iodide burn complexes.

In a parallel study, no effects of silver iodide from burn complexes were observed on the caecal flora and their function in rabbits (Jones and Bailey, 1974). Rabbits eliminated essentially all of an oral dose of silver iodide in 6.3 days on the average, and 99% of labeled silver iodide was eliminated in 3 days. Only 8 to 25% of silver was found to enter the caecum, and no accumulation was observed. Dry matter disappearance and folic acid production were not reduced with free silver ion or iodide forms, when used at levels above those which might be expected in ingested materials.

Thus, the short retention times of plant materials containing seeding agents, and the high chloride and organic matter concentrations, render these compounds of minimal impact in the digestive systems of ruminants and caecum animals.

13.5 EFFECT ON BIRDS

There is limited information available on possible effects of silver ion on birds, and essentially no information on possible effects of silver iodide or silver burn complexes.

It seems unlikely that silver in various forms would have greater effects on avian physiological processes than on humans or other mammals, with the possible exception of egg production and eggshell strength. On a more general biochemical level, silver can be antagonistic to copper metabolism in cattle and chickens, predominantly in chickens raised with copper-deficient diet regimes (Davis, 1972). In a similar experiment, silver ion reduced the toxic effects of higher selenium levels, while with excess silver, a selenium deficiency in chickens was noted (Bunyan et al., 1968; Ganther et al., 1973).

In research carried out with chicks, Hill et al. (1964) found that silver was antagonistic to copper, in that all copper deficiency symptoms were accentuated in the presence of silver and absence of copper (Table 13-4). Of special interest was the ineffectiveness of silver when sufficient dietary copper was present. They attributed these effects to basic similarities between silver ion and copper ion, and to the dissimilarity of either of these to mercury, thus precluding antagonism between silver and mercury.

In spite of this postulated similarity between copper and silver based on ionic measurements, Jensen (1974) noted that dietary silver and copper modify selenium metabolism in the chick by different mechanisms, as revealed by experiments in which combinations of these metals at various concentrations were fed as soluble salts.

These results suggest that in nutritionally stressed avians, silver presence may affect copper utilization, but relatively higher levels of ionic silver might be required.

In these studies of silver-copper interaction effects on chickens, no mention has been noted of specific problems with hatchability or strength of eggs.

13.6 INSECTS AND EFFECTS OF METALS

Available information concerning possible silver effects on insect physiological characteristics and activity are not very relevant to studies of

TABLE 13-4. In vivo *interaction between silver and copper on body weight and mortality in chicks (Hill et al., 1964).*

	Cu. μg/g			Cu. μg/g		
	0	10	25	0	10	25
Ag	3 week body wt[2]			Mortality, 4 week[2]		
μg/g	g	g	g			
0	145	196	232	5	2	3
10	125	190	229	5	1	0
25	139	199	244	3	0	2
50	126	240	243	10	2	0
100	110	174	240	12	1	4
200	96	187	207	13	0	0

[1] Twenty chicks started in each treatment.
[2] Mean of 4 determinations in each treatment.

weather modification, as silver ion has again been the predominant form used. In most of this research, the specific ions and the chemical composition of the natural waters used were not described, making it difficult to establish the effects of chemical equilibria on silver ion availability.

Warnick and Bell (1969), in a general review, noted previous studies describing *Daphnia magna* sensitivity to silver ion in hard river water, at a threshhold toxicity level of 0.03 mg/liter. In Lake Erie water, the threshold for immobilization was reported to be 0.005 mg/liter of silver ion (Anderson, 1948), a level which might be found in natural environments under specific conditions. A 15-day LC_{50} of 8.8 μg/liter silver was established with the stonefly (*Pteronarcys californica*). Five other acute toxicity tests did not yield LC_{50} values because of greater than 50% mortality in the lowest test concentrations of silver. The lowest silver concentration tested was 4 μg/liter, which killed 80% of the 24 mm *P. california* in 7 days. Younger specimens were found to be more susceptible than larger instars near their final molt. High correlation coefficients, greater than 0.89, were demonstrated between the silver concentration in the insects' bodies and the concentration in water.

Studies using nematodes provide insights into possible effects of ionic silver in terrestrial environments. Pitcher and McNamara (1972) conducted *in vitro* tests of silver ion toxicity to plant-parasitic nematodes. *Pratylenchus penetrans* was most susceptible to silver, showing 99 and 100% lethality at concentrations of 0.01 and 0.1 μg/g, respectively. Silver ion solutions of 0.1 μg/g killed no less than 96% of nematodes tested, while 0.01 μg/g killed most *P. penetrans*, 42% of *X. diversicaudatum*, and 25% of *A. ritzemabosi*. The authors noted, however, that using silver ion as a controlling agent for nematodes was not advisable due to the high cost of silver salts, and to potential phytotoxicity.

13.7 CONCLUSIONS

Available information about potential silver effects on higher terrestrial organisms does not relate directly to weather modification program assessment, as silver ion was used in most of these investigations. However, information on possible effects of silver iodide and burn agent complexes on rumen and caecum functions of animals is available.

Silver has been found to be an antagonist for copper, both in rats and in chickens, and studies of excretion processes show that silver is largely excreted following patterns observed in humans.

Studies of possible silver effects on tumor development and regression also relate only indirectly to weather modification problems, as the forms, levels, and contact times bear little resemblance to silver exposures in the environment.

The possible effects of silver iodide and burn agent complexes on rumen and caecum functioning in particular animals have been studied; with the short retention times and the complexing ions present in these environments, no effects of nucleating agents ingested with plant materials were noted.

A wide range of terrestrial organisms has been investigated for silver effects, using silver ion, but it is difficult to predict potential silver iodide effects based on this information. However, with the low levels of free ionic silver released from the halide complexes, and the reaction of the silver cation with a wide range of organic and inorganic materials in the natural environment, the effects of silver derived from weather modification activities are expected to be extremely difficult to detect or monitor, should they occur.

14

Other Seeding Agents: Alternatives to Silver Iodide

W. R. Barchet

14.1 INTRODUCTION

Seeding agents are applied to clouds to generate large water droplets by stimulating coalescence, or to catalyze the formation of ice crystals within supercooled clouds. Seeding clouds with silver iodide is intended to cause the latter. This chapter deals with materials other than silver iodide which also serve to nucleate ice crystals within supercooled clouds. Some of these materials may provide realistic alternatives to silver iodide seeding for research and operational weather modification programs. Agents which show promise as alternatives to silver iodide fall into four broad categories:

(1) organic crystalline ice nucleants,
(2) other inorganic crystalline ice nucleants,
(3) chilling agents, and
(4) biological ice nucleants.

Much of the present dominance of cloud seeding programs by silver iodide can be traced to the development of a very effective technology for the delivery of this seeding agent to the cloud. Early laboratory studies demonstrated the ease with which an aerosol with a high ice nucleation activity could be produced. Thus, delivery systems were rapidly developed. Today the silver iodide aerosol, produced by any one of several effective generation techniques, serves as the standard against which all other potential ice crystal catalysts are eventually compared.

All operationally successful ice catalysis agents must demonstrate: (1) the ability to catalyze ice crystal formation, (2) the ability to be efficiently dispersed and delivered as an aerosol, (3) cost effective ice crystal production, and (4) an acceptably low ecological impact. An evaluation of which of several agents is better must be based on the overall ability of the agents to meet these requirements.

Effective ice catalysts generally possess a high onset temperature for ice crystal formation, i.e., -5°C or higher. However, the entire temperature spectrum of ice nucleation activity is important since it is the production of ice crystals at temperatures below the onset temperature that is important in cloud seeding. In many applications a rapid increase in activity to a high per mass plateau is desirable so that most crystals are nucleated at the highest possible temperature and an abundance of crystals is not created at the very cold temperature where natural nuclei become effective.

Aerosol generation, agent dispersal, and/or agent delivery present the greatest obstacles to the operational acceptance of many potential seeding agents. For agents acting as ice nuclei, maximum efficiency is achieved when they are dispersed as an aerosol of small particles. Mechanical dispersal by grinding is always ineffective, except for dry ice dispersal. The most widely used method is to vaporize the material into a hot carrier gas and to rapidly quench the vapor to bring about the condensation of the agent into small crystalline particles. For this process to be effective, the material must be volatile at moderately high temperatures and resist thermal decomposition and oxidation of its active surface. In addition, the size spectrum of the particles produced must be such that few particles are lost before reaching the activation zone in the cloud through evaporation if the agent is too volatile. The material must also be sufficiently insoluble so that particles are not lost through dissolution in cloud drops at temperatures above the onset of freezing activity of the material. The particles must be small enough to make the economy of seeding large. Although small sizes are necessary to obtain high rates of contact nucleation, larger particles may be effective in deposition nucleation of ice crystals.

Not all cloud seeding objectives or opportunities call for identical seeding techniques. Generator and dispersal systems must be capable of delivering the seeding agents where needed. Methods of seeding agent dispersal adequate for cloud-base seeding may be impractical for on-top or in-cloud seeding. Wherever the seeding agent is to be delivered, operational utility requires that the dispersal system be simple, rugged, reliable, and inexpensive.

Although the cost of the seeding agent is usually a small fraction of the total cost of an operational cloud seeding program, the economics of seeding agent performance can not be overlooked. To compare various agents requires an evaluation which includes the cost per ice crystal produced, the cost of constructing, maintaining, and operating the delivery system, the cost of storing and transporting the agent, and the cost of protecting personnel and equipment from the effects of the seeding agent.

Of equal importance to the technological and economical suitability of a seeding agent is its ecological acceptability. For personnel involved in the handling and dispersal of the agent, acute and chronic toxicity, carcinogenicity, and mutagenicity of the agent or its by-products are important considerations; no adverse human response to the agent is highly desirable. Furthermore, when dispersed into the environment, the agent and its by-products should be incapable of bioconcentration in any segment of the escosphere and should degrade rapidly into materials compatible with the ecosystem (Cooper and Jolly, 1970).

14.2 ORGANIC CRYSTALLINE ICE NUCLEANTS

Bashkirova and Kariskov (1957) initiated the search for organic ice nucleants with the discovery that phloroglucinol dihydrate crystals nucleated ice crystals at temperatures as warm as silver iodide. By 1967, well over 300 organic compounds had been screened for ice nucleation activity, and more than 50 demonstrated the ability to nucleate ice under various conditions at temperatures warmer than -5°C.

Surveys of numerous compounds were carried out by several investigators. Komabaysi and Ikebe (1961) found ice nucleation-active aromatic compounds using a mixing cloud chamber. Using cold-stage microscopy, Head (1961, 1962a,b) discovered ice nucleation activity in 7 of 30 steroids, in several fluorenes, and identified phenazine as a material also active as a sublimed smoke in a mixing cloud chamber. Several amino acids were also found active (Power and Power, 1962). Optically active and racemic forms of some amino acid crystals showed different onset temperatures for ice nucleation activity (Barthakur and Maybank, 1963, 1965; Parungo and Lodge, 1967). Langer et al. (1963) surveyed some 35 compounds but found only phloroglucinol to be promising. After Fukuta and Mason (1963) found evidence of expitaxial growth of ice crystals on several organics, Fukata (1963) reported on the extremely high onset temperature of -0.6°C for metaldehyde crystals on the cold stage and the efficient production of ice crystals by sublimed metaldehyde smokes at -10°C in a mixing cloud chamber. In a survey of solvents that promoted ice nucleation on organic compounds by evaporative cooling, Fukuta (1965a,b) introduced 1,5-naphthalenediol as an effective organic ice nucleant. The last major search for organic ice nuclei was undertaken by Fukuta (1966) and involved some 300 compounds of which 25 were found active at or above -5°C; 6 compounds had nucleation threshold temperatures nearly as warm as metaldehyde.

The chemicals identified in the surveys cited above have a wide variety of crystal and molecular structure. Ice nucleation activity seemed to be the only unifying characteristic. The concept of epitaxial nucleation of ice crystals by matching the ice lattice with a foreign crystal lattice was widely used to explain why some materials were active and others failed to work. However, some active ice nucleants were crystallographically quite unlike ice, so epitaxy alone could not explain why they were effective.

Nearly all of the active materials have the ability to form hydrogen bonds with water. The interaction between water molecules and the surface of the nucleant through hydrogen bonding was used by Garten and Head (1964) to identify 1,3,5-cyclohexanetriol as an ice nucleant. As the hydrogen bonding ability of a series of phenols and benzoic acids increased, Parungo and Lodge (1965) found that the onset temperatures for ice nucleation also increased. Molecular structure appeared to have a greater effect on onset temperature than the crystallographic configuration. After examining a large number of organic compounds, Garten and Head (1965) described how the distribution and strength of hydrogen bonding groups (HBG's) on the nucleant surface controls the ice nucleation behavior of the material.

Good ice nucleants (Garten and Head, 1965) were found to: (1) expose single HBG's at the same depth with a spacing that is a multiple of the ice lattice with a disregistry of less than 7%, providing for classical epitaxial nucleation and growth (phenazine and fluorenone-I), (2) have HBG's in a hexagonal pattern with a close fit to the 001 ice plane (phloroglucinol dihydrate and 1,3,5-cyclohexanetriol), (3) provide small clusters of HBG's in which the spacing of hydrogen bonding sites is nearly the same as the oxygens in a close packed ice plane (metaldehyde), and (4) have HBG's randomly distributed but with an optimal surface density somewhat less than the 5-7 per $100A^2$ for ice (graphites). Hydrogen bonding group patterns characteristic of poor ice nucleants were associated with large disregistries, HBG densities in excess of that for ice, HBG's inaccessible because of surface configuration, or HBG's unavailable at the surface.

The hydrogen bonding concept can explain why some organic crystals have ice nucleation threshold temperatures above -5°C. However, a high onset temperature alone is not sufficient to make the material competitive with silver iodide. Nucleating ice on a large organic crystal immersed in water is not the same as nucleating ice on an aerosol of the material in a cloud chamber or in a real cloud. To be competitive with silver iodide, the material must be able to yield an aerosol which, on a cost per active nucleating particle, is comparable to the activity of silver iodide.

Early attempts at organic nucleant aerosol generation relied almost wholly on sublimation at high temperature, followed by rapid quenching to initiate the condensation of the organic vapor into small particles. Many of the materials decomposed at the high temperatures needed to create a smoke or they oxidized rapidly, losing their effectiveness as ice nucleants.

For other materials, the rapid quenching led to glassy rather than crystalline particles, with an attendant loss of activity (Fukuta, 1966).

Most of the organic ice nucleants are relatively insoluble. However, some are sufficiently soluble that most small particles could dissolve should they be captured by cloud drops above their onset temperature for ice nucleation. Evans (1966), however, found that some amino acid crystals had a higher onset temperature if the surface was etched by dissolution as nucleation occurred. In contrast, Fukuta (1975) found contact nucleation to be more active in terms of nucleation threshold than freezing nucleation for some organic aerosols.

Of the more than 300 organic compounds tested for ice nucleation activity, approximately 50 had onset temperatures in excess of -5°C. Thermal instability, poor aerosol formation, and high solubility reduced the number of materials which appear promising to the following five compounds: 1,3,5-cyclohexanetriol, metaldehyde, 1,5-naphthalenediol, phenazine, and phloroglucinol dihydrate. These materials had shown adequate activity as ice nucleants in the laboratory to arouse interest in applying these materials to real clouds.

Phloroglucitol (1,3,5-cyclohexanetriol) smoke prepared by sublimation produced some 10^{11} nuclei/g at -12°C in the laboratory of Garten and Head (1965). The alicyclic and polyalcoholic nature of this compound led them to regard it as nontoxic; no other data exist to substantiate this assessment. Although this material appeared promising in laboratory tests, there is no evidence of any attempt to develop large scale aerosol generation techniques for phloroglucitol.

Phenazine, as a sublimed smoke, produced some 10^{11} nuclei/g at -12°C in a mixing cloud chamber. However, because of its potential toxicity, Garten and Head (1965) concluded that its use as a seeding agent would be hazardous. No development of aerosol generation techniques have been reported for phenazine since these laboratory studies.

Only three compounds, metaldehyde, 1,5-naphthalenediol, and phloroglucinol dihydrate, remain as promising organic ice nucleants. Dispersal of these agents in an efficient and effective manner is still the major technical obstacle to acceptance of these materials as alternatives to silver iodide seeding. Fukuta (1972b) presented a thorough review of the methods by which organic substances can be dispersed as fine particles suitable for seeding clouds; a short summary of his findings follows. Small particles can be produced by two methods: dispersal of the solid into smaller particles, and by condensation from the vapor phase. Rate of particle production, activity of the resulting particles, and size of the particles attained rule out the use of physical dispersal methods in operational applications, although claims have been made to the contrary (Knollenberg, 1966b).

Generation of ice nucleus aerosols by vapor condensation methods is among the earliest described for organic materials. A small amount of material is sublimed into a hot carrier gas which is then thermally quenched to produce a high supersaturation of the carrier with the organic vapor. Condensation by either homogeneous or heterogeneous nucleation of organic particles within the vapor produces the ice nucleus smoke. The particle size distribution is controlled by the rate at which particles are nucleated, the rate which these particles consume the vapor, and the amount of vapor available for growth. Both rates are sensitive to the supersaturation achieved on quenching the vapor. To produce a large number of small particles, the supersaturation should be as large as possible but maintained for only a brief period to prevent excessive growth of the particles after they are formed.

Rapid quenching and high supersaturation can be realized by two methods: jet mixing and supersonic isentropic expansion. In the jet mixing approach a jet of hot, vapor laden air is extruded into a cold environment. High supersaturation is achieved in a small zone in which thermal quenching has cooled the carrier gas but mixing with the environment has not yet diluted the vapor. The isentropic expansion of the vapor laden carrier gas through a supersonic nozzle provides an internal cooling mechanism that avoids diluting the vapor during quenching. Both approaches are capable of producing an aerosol with most particles of submicrometer size.

The earliest large scale organic smoke generator used the jet mixing principle and produced metaldehyde smoke with 10^{12} nuclei/g active at -12°C (Fukuta et al., 1966). In this device, pellets of metaldehyde were evaporated by electrically heated air and discharged into cold air through a low pressure jet. Particle diameter averaged around 0.5 μm. An airborne version of the jet mixing device used the heat from the aircraft engine exhaust to vaporize a continuous feed of fluidized metaldehyde powder (Fukuta, 1967). Improvements in ice nucleus production with metaldehyde were achieved by suspending the powder in water and vaporizing the mixture prior to quenching (Fukuta, 1968); this technique further simplified the handling of this agent, 1,5-naphthalenediol or phloroglucinol. The supersonic nozzle approach was described by Fukuta (1972b) using 1,5-napththalenediol in superheated steam at 200-300°C and 5 atm. A ground based smoke generator for the three organics using superheated steam and a supersonic nozzle produced copious numbers of submicrometer particles (Fukuta, 1972b). Refinements on the superheated steam-supersonic nozzle method were reported by Fukuta et al. (1973) and culminated in a U.S. Patent to Fukuta and Paik (1974) on generating ice nucleus smoke particles by this technique. Development of improved airborne organic ice nucleus generators (Armstrong et al., 1974) continues under the supervision of Dr. N. Fukuta.

The ice nucleation characteristics of organic aerosols appear to depend on the method of aerosol generation and on the mode of ice nucleation. Organic aerosols appear to show higher threshold temperatures as contact nuclei than as freezing catalysts (Fukuta, 1972b, 1975). Metaldehyde freezing activity begins at a temperature some 2°C colder than its contact activity. Even greater retardation of freezing activity is seen for 1,5-naphthalenediol (4°C) and phloroglucinol ($>$5°C). For some materials the carrier gas has an effect on the spectrum of ice nucleation activity. Metaldehyde responds to quenching in the presence of water vapor with a slight decrease in onset temperature followed by a rapid rise to a plateau of 10^{12} nuclei/g at -12°C. When prepared by jet mixing from a dry powder, its activity near onset is greater but the output is only 10^{8} nuclei/g at -12°C. A similar response is noted for phloroglucinol but not for 1,5-naphthalenediol. At temperatures much lower than the onset value the presence of water vapor appears to increase the number of nuclei whereas near the onset temperature smokes produced from the dry powder are more active (Fukuta, 1972b).

When released into the atmosphere, the activity of organic ice nucleants may be reduced by several processes. Although several investigators noted that some organics are more resistant to photo-deactivation than silver iodide, Garten and Head (1965) found that photo-oxidation of the surface may destroy the ice nucleation activity of some organics. Surface adsorption of impurities and chemical decomposition may limit the useful life of organic aerosols (Fukuta, 1972b). Volatility, solubility, and coagulation also may reduce the number of active particles. The length of time during which an ice nucleus aerosol remains active is important to the operational use of a particular material. The high vapor pressure of metaldehyde causes metaldehyde smokes to quickly evaporate. Rapid coagulation of dense metaldehyde smokes is promoted by the large electric dipole of the metaldehyde crystal. Thus metaldehyde aerosols should be released within the supercooled portion of a cloud for maximum effect (Fukuta et al., 1966). Phloroglucinol and 1,5-naphthalenediol are sufficiently nonvolatile and insoluble that cloud base or ground level seeding may be practical with these agents.

The limited active life time of organic ice nuclei in the atmosphere can be considered an advantage in many situations. Once the material evaporates or deteriorates there is little chance of its reactivation. While extra area effects due to downwind transport of nuclei will be minimized, other effects of seeding due to cirrus ice, hydraulic jump effects, enhanced vertical transport, reduced pressure fields, etc., will not be diminished.

Field trials of organic ice nuclei have been very limited. Those which have been conducted were of an exploratory rather than systematic nature. Braham (1963) conducted the earliest test of seeding natural clouds with an organic ice nucleant. Phloroglucinol dihydrate was released into the upper portion of stratoform clouds from an airplane. Releases of phloroglucinol in an alcohol slurry and as a dry, milled powder were more effective than the untreated, commercial powder. But the snow showers from even the most successful release were less intense than even the weakest produced by a minimal release of dry ice.

The earliest field trials with metaldehyde were carried out by Fukuta (1967, 1972a) in Flagstaff, Arizona on four days in August 1966. Isolated cumuli were seeded from an aircraft during penetration and at cloud base. While two seeded clouds developed a radar echo and precipitation area, the five other treated clouds showed no significant response.

Summertime cumulus clouds were again the target for seeding with organic ice nuclei near Rapid City, South Dakota during August 1974 (Armstrong et al., 1974). An improved smoke generator for 1,5-naphthalenediol was used to seed the clouds during penetration at the -5°C level. On only four days did the cloud liquid water content satisfy seedability requirements. All six seeded clouds glaciated within 25 minutes after seeding. Virga was observed to descend from four of the six seeded clouds, but not from other clouds in the area.

These very limited tests of organic ice nucleus seeding can not be used to make a strong case for or against the efficacy of seeding with organic compounds. However, all of the other evidence such as aerosol ice nucleation activity, aerosol behavior, and improved generator technology, seems to give cause for a broader examination of the seeding potential of organic ice nucleants than has thus far been accomplished.

Organic ice nucleants presently are attractively priced compared to silver iodide. Armstrong et al. (1974) gives the bulk cost ratio against silver iodide as 0.008, 0.03, and 0.07 for metaldehyde, 1,5-naphthalenediol, and phloroglucinol, respectively. At temperatures above -10 to -8°C, the cost per crystal produced shifts heavily toward the organic ice nucleants which exhibit higher onset temperatures than silver iodide. Present generator systems for the organic nucleants appear well suited to cloud base and cloud penetrative seeding by aircraft, and do not seem to require greater mechanical expense than silver iodide acetone solution burners.

The hazards associated with organic ice nucleants fall into two areas: (1) acute toxicity and flammability hazards associated with the storage, preparation, and delivery of the agent, and (2) the long term, chronic toxicity and ecological impact of the use of these agents in seeding programs. In both areas only limited data are available to serve as guides to assessing the hazard potential of these materials.

Exposure of personnel to the materials during storage offers minimal hazard; metaldehyde, 1,5-naphthalenediol, and phloroglucinol are solids and can be purchased in metal drums. Fukuta (1972b) and Armstrong et al. (1974) claim water suspensions of these organics pose no flammability hazard to inflight transport of the agents. Delivery of the agent by aircraft should not expose flight personnel to hazardous concentrations of the agents. However, for ground based delivery systems, some regions within the nucleus plume may exceed tolerable levels.

Standard toxicology references, such as the Merck Index (8th ed., 1968) or Sax (1975), give only a rough estimate of the potential hazard associated with these chemicals. Ingestion of metaldehyde or phloroglucinol in unspecified but certainly above trace quantities may produce liver and kidney damage (Verschuuren et al., 1975). Metaldehyde poisoning is a hazard to pets which ingest molluscide pellets containing 2 to 3% of this agent (Udall, 1973). Yet metaldehyde is a permissible food additive, albeit in small amounts, according to Federal Drug Administration code (Sax, 1975). Intraveneous administration of 1,5-naphthalenediol at 1.9 millimoles Kg^{-1} or phloroglucinol at more than 15 millimoles Kg^{-1} to mice was found to induce convulsions in 50% of the mice tested (Angel and Rogers, 1972). However, it is unlikely that the materials will be ingested or injected during routine handling. Absorption through the skin causes phloroglucinol to be rated as moderately toxic (Sax, 1975). Standard, normal hygenic and safety precautions should avoid any hazard associated with an acute exposure to these materials.

The effect of long term, low concentration exposure with the operational use of these materials is more difficult to assess. In this regard, the biodegradability of the organics is an advantage, for accumulation in the biosphere is not likely to occur. Soviet industrial hygienists suggest 0.2 mg/mg^3 metaldehyde as the maximum allowable concentration in the air of the working zone (Dobryanskii, 1974). This value would be exceeded only very near to a ground based generator. Similar standards for the other organic agents have not be established.

Only a few studies shed light on the impact of organic ice nucleants on the biosphere. Phloroglucinol was found to stimulate carrot growth at concentrations less than 150 $\mu g/g$ in soil while concentrations exceeding 175 $\mu g/g$ retarded growth (Meskhi and Dzhokhadze, 1973). Some rumen bacteria are capable of anaerobic degradation of this agent but the tolerable concentrations were not specified (Tsai and Jones, 1975). The long term pollution potential of metaldehyde is likely to be minimal because depolymerization and oxidation to acetic acid (Garten and Head, 1965) will remove it from the biosphere. Also, some mammilian systems metabolize metaldehyde by the same pathway as alcohol and thus should be able to tolerate the very low levels of metaldehyde ingested along with forage crops watered by rain from metaldehyde seeded clouds. However, the behavior of 1,5-naphthalenediol or phloroglucinol at low concentration in the biosphere is not well documented.

The only study which has specifically examined the ecological impact of organic seeding agents was by Church et al. (1975). They studied metaldehyde, 1,5-naphthalenediol, phloroglucinol, and silver iodide in aquatic and soil systems and observed the effects of these materials on natural biota and fish in aquaria, and respiration rates and microorganisms in soils. All seeding agents were at a concentration of 1 mg $liter^{-1}$ pond water in the aquaria and at 1 mg/g in the soil tests.

In the aquaria study the organic agents stimulated and maintained an increased level of bacterial growth. Silver iodide stimulated but did not sustain a high growth rate. Metaldehyde and 1,5-naphthalenediol had no effect on algae, protozoa, or fish populations. However, phloroglucinol was slightly toxic to algae and to fish feeding on algae. Silver iodide was also slightly toxic to algae and inhibited the growth of fish feeding on algae as well as reducing fish reproduction efficiency.

Soils from three locations: arid nonagricultural soil from Denver, Colorado; conifer forest soil from Helena, Montana; and rich garden soil from Nonjenay, Maryland were used to demonstrate soil respiration and bacterial responses to the seeding agents. Organic seeding agents affected a general stimulation of bacterial growth and oxygen uptake. The effects were more pronounced in soils of high organic matter content. Silver iodide slightly depressed the total bacterial flora with the nitrogen fixing organisms in rich soils being most inhibited.

Degradation studies were inconclusive. The seeding agents became undetectable in raw soil in a shorter time than in sterile soils. However, cultures of soil bacteria on artificial media to which the organic agents were added failed to degrade the seeding agent.

These studies by Church et al. (1975), summarized above, are a first step toward demonstrating the ecological acceptability of organic seeding agents. At the concentrations used, which are most likely high in comparison to the levels that would be produced by seeding, the organic agents have not engendered any hazardous responses in the aquatic and soil systems studied. However, the possibility for bioconcentration of either the agents or their by-products has not yet been adequately explored.

14.3 INORGANIC CRYSTALLINE ICE NUCLEANTS

The discovery by Vonnegut (1947) that silver iodide and lead iodide were capable of nucleating ice in supercooled clouds began a major search for inorganic ice nuclei. Materials were sought which closely matched the crystallographic parameters for ice, were insoluble or slightly soluble in water, and were thermally stable. Hosler (1951) identified Ag_2S as an active nucleant in a mixing cloud chamber. Mossop (1956), using an expansion chamber, confirmed the activity of PbI_2. Pruppacher and Sanger (1955) added CuS, Ag_2O, and V_2O_5 to the list although iodine vapors may have contaminated some of their tests. Mason and Hallet (1956, 1957) confirmed the activity of CuS. Katz (1960) demonstrated the activity of CuO as well as Cu_2O. Mason and Van den Hauvel (1959) reviewed the relationship between the crystallographic properties and ice nucleation activity of inorganic compounds. In general, insoluble substances which closely matched the ice crystal lattice had the highest onset temperature for ice nucleation; the materials mentioned above produced at least one ice crystal for 10^4 particles at temperatures warmer than -8°C.

Since the early 1960's very little additional work has been conducted in search of new inorganic ice nucleants. Most attention was focused on mechanisms of silver iodide and lead iodide activity in clouds. In a study of water adsorption by ice nucleants, Zettlemoyer et al. (1963) found that heat treatment of ordinary silica gel produced an active ice nucleating powder when the surface density of hydrophilic sites was optimized. However, this material could not be dispersed in a form effective for cloud seeding. Improvements in silver iodide smoke generation involved altering the chemistry of the solution or pyrotechnics burned. Solutions in which the silver iodide is complexed with NH_4I and pyrotechnics using silver iodate presently dominate the silver iodide seeding technology.

Passarelli et al. (1973, 1974) carried the concept of a close match between an artificial ice nucleant crystal lattice and the ice lattice to a logical conclusion by producing a solid solution of AgI and CuI with essentially zero disregistry to the ice basal plane. An aerosol generated by nebulizing a solution of CuI-3AgI dissolved in a 30% hydroiodic acid solution produced slightly more ice crystals than a similarly dispersed pure AgI solution. At the onset temperature for nucleation, the activity of the CuI-3AgI was up to ten times greater than that of pure AgI. In bulk, the CuI-3AgI ice nucleation onset temperature was more than one degree Celcius warmer than that for silver iodide.

Atomization of a solution containing the appropriate mixture yields particles of the correct composition. However, the process is relatively slow and at present is not a technique likely to be adaptable for field use. Aerosols of the correct stoichiometry are also produced when the solid solution is evaporated in a flame and condensed. Thus it appears feasible to use CuI-3AgI in standard AgI-NH_4I solution burners but no studies are available to support this transfer of technology. The use of CuI as part of the seeding material reduces the amount of silver used by about 25%.

An innovation also aimed at reducing the amount of silver needed to produce an active aerosol was put forth by Corrin (1976). Tetrabutyl-o-titanate was added to a weak AgI-NH_4I in acetone solution. When this solution was burned, the refractory material TiO_2 was the first to crystallize from the hot vapor. The small particles of TiO_2 served as a substrate for the subsequent deposition of a thin layer of AgI as the temperature of the burner exhaust fell to where this material began to condense. The activity of the TiO_2-based aerosol rivaled that of the standard AgI-NH_4I solution with a 90% reduction of the amount of silver iodide used in preparing the test solution. The future of this method in operational programs is unknown.

Soviet scientists have experimented with metal chelates as ice nucleants and have found a mixture of β-diketone and β-ketoester copper chelates to be effective at temperatures as warm as -1.5°C (Patrikeev et al., 1974). Mixtures of copper acetylacetonate (25-50%) and copper derivatives of ethyl acetylacetate (75-50%) had the highest activity. Battan (1977) reported that the Soviets are very interested in these compounds for operational use. Ice nucleation activity of copper acetylacetonate as high as $10^{13}g^{-1}$ can be produced by flame, electric arc, or pyrotechnic vaporization of the very stable copper chelate (Battan, 1977). Copper acetylacetonate [also copper *bis* (2,4-pentanedionate - 0,0')] is a planar β-diketone chelate (Martell and Calvin, 1952) which forms blue crystals with a melting point of 230°C. Sax (1975) considers this substance to be relatively nontoxic although this and other related chelates have been used as fungicides (Popelka et al., 1973). Little additional information is available on the ice nucleation or ecological behavior of these compounds.

Intensive development of particle generating techniques for the inorganic ice nucleants occurred only for silver iodide, with some parallel development for PbI_2. As with organic ice nucleants, the lack of proven dispersal and delivery system is the major obstacle to using CuI-3AgI, or the TiO_2-based AgI nucleants operationally. Much of the current technology for generating AgI aerosol appears transferable to inorganic ice nucleants. However, because CuS reverts to CuO and SO_4 at pyroburn temperatures, no practical dispersal technique is available for this material.

Most of the inorganic ice nucleants discussed offer significant economic advantages over silver iodide. The TiO_2-based AgI smoke could reduce silver iodide consumption 90% and the cost of the tetrabutyl-o-titanate and silver iodide solution is only 28% that of a normal silver iodide in acetone solution (Corrin, 1976). The solid solution CuI-3AgI substitutes copper for 25% of the silver to reduce the bulk cost of the mixture by 20% from that for silver iodide alone. Seeding apparatus for these materials would probably require only slight modifications of the generators presently used to dispense silver iodide aerosols.

The ecological impact of these materials as the seeding agent in an operational weather modification program is largely unknown. Agents containing AgI would behave in ways discussed elsewhere in this book except that less silver iodide would be added to the ecosphere. However, since both silver and copper are known bactericides, the impact of any synergistic effects would have to be carefully monitored before undertaking large scale testing of the CuI-3AgI nucleant.

14.4 CHILLING AGENTS

Chilling agents promote ice nucleation by cooling a small volume of air to a sufficiently low temperature so that a large number of natural ice nuclei can be activated or so that ice will be homogeneously nucleated within the air or within small drops cooled along with the air. Several processes can achieve the required chilling: sublimation of solid carbon dioxide (Schaefer, 1946), evaporation of volatile liquids (Schaefer, 1946; Serpolay, 1954), adiabatic expansion of gases (Vonnegut, 1948), and dissolution of substances with large endothermic heats of solution (Knollenberg, 1966).

14.4.1 Dry Ice

Schaefer (1946) discovered that ice crystals are produced when a very cold object is passed through a supercooled cloud. If the object was colder than about -40°C a large number of crystals were produced; objects at higher temperatures produced fewer crystals. The most effective source of cold objects was a piece of dry ice scratched above the cloud. The first dry ice seeding of supercooled stratus clouds was carried out by Vincent Schaefer at Mt. Greylock, Massachussetts on November 13, 1946. Many other field tests of dry ice seeding were performed after this demonstration of its effectiveness.

Predictions by Langmuir (1948) and others of the crystal production per mass of dry ice sublimed gave numbers larger than observed. Several experimental investigations of ice crystal production by dry ice have been made (e.g., Eddie and Mee, 1963), with the recent study by Fukuta et al. (1971) being the most reliable. Fukuta et al. measured the number of crystals produced as a dry ice pellet of known size fell through a given depth of cloud. From measurements and estimates of the evaporation rate and fall speed of the pellet they deduced a crystal production rate near 10^{12} g^{-1} evaporated and found the production nearly independent of cloud temperature below -0.5°C and of the fall speed of the pellet. This constancy of ice crystal production over a wide range of temperature is significant to the cloud seeder; he knows precisely how many crystals he will produce for a given release of seeding agent provided all the dry ice evaporates in the supercooled portion of the cloud.

Clouds may be seeded with dry ice by dropping the material into the cloud; Schaefer (1976) and Holroyd et al. (to be published) make strong arguments for the use of dry ice in on-top seeding of cumulus clouds. Also, if the temperature at cloud base is below 0°C, dry ice seeding beneath cloud base can produce crystals which may be lifted into the cloud to induce effects similar to an in-cloud release of the material. Surface seeding with dry ice has been used to clear supercooled fogs (e.g., Gaivoronskii et al., 1968). On-top seeding presents some interesting and as yet unexplored problems. Although ice crystal production is constant over a wide range of temperature per gram of dry ice evaporated, the cloud seeder may wish to stimulate crystal formation at temperatures between -5°C and 0°C and not cause the cloud to glaciate at lower temperatures. The release of a dry ice pellet into the top of a cloud may well overseed the upper region of the cloud, and if the updrafts are strong enough, the pellet may not even reach the desired seeding level.

Most field trials of dry ice seeding have involved stratoform clouds or fogs of low liquid water contents. When a pellet of dry ice of sufficient size to survive the trip from cloud top to the O°C level is released in a cumulus cloud of high liquid water content, collisions with supercooled drops are bound to occur and will most likely cause the rapid riming of the dry ice pellet. At present, the consequences of dry ice pellet riming have not been investigated theoretically or experimentally. However, it is quite reasonable to assume riming will change the rate of heat transfer between the pellet and air and through the latent heat of water fusion will deplete the mass of the pellet with a drastic alteration of the production of ice crystals in the air.

Such problems are, however, problems of engineering. Dry ice is relatively easy to transport or to manufacture on board an aircraft from carbon dioxide gas or liquid in high pressure cylinders. Apparatus for crushing and dispensing dry ice has been described (e.g., Appleman, 1968). The technological requirements for dry ice delivery systems do not appear unmanageable.

The economics of dry ice seeding in comparison to silver iodide seeding are difficult to assess. Per unit mass, the cost of liquid or solid carbon dioxide is about 3×10^{-4} that of silver iodide. However, the per ice crystal produced cost ratio will depend on the seeding temperature and on how much dry ice must initially be released to seed the cloud at that temperature. From the activities of dry ice and AgI, a conservative estimate would place the amount of dry ice needed to seed a cloud at -10°C at ten times the amount of silver iodide needed for the same effect. Losses of dry ice enroute to the seeding target would increase this factor. In only slightly supercooled stratoform clouds, the dry ice pellet size could be engineered to provide complete evaporation within the supercooled zone to achieve maximum benefit. In cumulus clouds of high liquid water content, the cost of delivery systems that avoid the difficulties mentioned above must be added to the cost of the seeding material.

Solid carbon dioxide seeding would have distinct ecological advantages over organic and inorganic ice nucleants. Carbon dioxide is a natural component of air; pure dry ice would leave no residue to contaminate the air or precipitation. Locally higher carbon dioxide concentrations may slightly decrease the pH of rain water but this effect is likely to be insignificant. However, some commercially produced dry ice may contain *Tris*-(hydroxyethyl)amine as a trace contaminant of the production process (Finnegan, private communication). Dry ice used in seeding projects, if free of potentially dangerous substances, should be environmentally acceptable as a seeding agent.

14.4.2 Volatile Liquids

The evaporation of a volatile liquid requires the transfer of vapor away from the liquid and a transfer of heat to the liquid to provide the heat of vaporization. Drops of highly volatile liquids will cool far below ambient air temperature in order to produce the heat transfer required for evaporation. Schaefer (1946) and Serpolay (1954) recognized that the chilling produced by evaporating liquid drops could catalyze ice formation in the same manner as dry ice.

Many liquids with high vapor pressure and latent heat of vaporization were tested as possible chilling agents. Serpolay (1960) introduced the use of liquid propane to clear supercooled fogs. Fukuta (1965a) found that organic ice nucleants had higher onset temperatures when dispersed as mixtures with certain volatile organic liquids; water soluble organic liquids did not give this effect. Further study led Fukuta (1965b) to a mixture of 40% chlorodifluoromethane (Freon 22) and 60% trichlorofluromethane (Freon 11) in a pressure pack as an effective chilling agent. Evaporative cooling of the drops had to exceed -40°C before ice crystal production became significant. In a laboratory study of the crystal production by several liquids, Fukuta et al. (1970) found propane and chlorodifluoromethane (Freon 22) produced 10^{11} to 10^{12} crystals/g^{-1} evaporated liquid. Recently Hicks and Vali (1973) described extensive laboratory and field tests which established the ice crystal production by liquid propane over a range of temperatures. At temperatures colder than -2°C, a constant production of about 4×10^{11}/g was achieved. This value was in good agreement with Fukuta's (1965b) theoretical estimate. Above -2°C the crystal production fell to near 10^{10}/g at temperatures within a few tenths of 0°C. Measurements by Andro and Serpolay (1970) gave somewhat lower values.

Because of its availability, liquid propane has been the most widely used liquid chilling agent for clearing supercooled fogs. Operational use for this purpose has been described (Serpolay (1965) in France; Hicks (1967), Gerdel (1968), and Fletcher (1971) in the United States). The use of liquid chilling agents appears limited to surface seeding of supercooled fogs because of the rapid evaporation of the small drops. New delivery methods must be developed for the remote dispensing of liquid chilling agents in the appropriate regions of cumulus systems if these clouds are to be seeded with these agents.

The widespread use of propane rather than silver iodide for supercooled fog clearing is partly due to the considerable economic advantage for propane but more so to the greater crystal production and hence effectiveness of propane near 0°C. Gerdel (1968) even suggests that clathrate formation may allow propane sprays to clear fogs up to +1°C.

Propane sprays apparently do not present a flammability hazard at distances farther than 3m from the nozzle (Hicks, 1967). An environmental impact study by Carroll (1970) also suggests that propane is not detrimental to the environment, nor does it react harmfully with constituents of the air. The suggestion by Hicks and Vali (1973), however, that propane is readily available and inexpensive may need to be revised for future cost estimates. Cryogenic liquids such as liquid nitrogen, oxygen, or air, are also effective chilling agents but are much more expensive at the present time than liquid propane and require specialized storage and handling. However, their use should be ecologically quite acceptable. In contrast, the use of the chlorofluoromethanes is not likely to be pursued because of the current question concerning the effect of these materials on the ozone layer.

14.4.3 Adiabatic Expansion of Gases

Production of ice crystals by adiabatic expansion to chill the air was first proposed by Vonnegut (1948). Although not put into operational practice for cloud seeding, the chilling and supersaturation that accompany adiabatic expansion provided a widely used means for detecting and measuring natural and artificial ice nucelus concentrations. The availability of other more effective, cheaper seeding agents kept this method from being further developed.

However, the recent controversy over the long range availability of petroleum products and the continual escalation of the cost of these products has again stirred interest in the use of compressed air as a source of ice crystals. Weinstein and Hicks (1976) describe laboratory and field tests of a supersonic nozzle compressed air expansion system capable of producing 10^8 crystals/cm^3 air at standard temperature and pressure at all temperatures below -4°C. The output of their system decreased to about 10^4/cm^3 at -2°C.

A direct cost comparison to silver iodide seeding is not possible. In this case the cost of producing ice crystals is related to the cost of producing the compressed air which depends on the amount of fuel used to drive an air compressor. By such an analysis, Weinstein and Hicks (1976) found that with their compressor more than 16 times more propane than gasoline is needed to produce the same number of ice crystals.

Ice crystal production by compressed air systems appears to be compatible with the ecosphere provided the pollution from the engine driving the air compressor is tolerable. Given the cost advantage over propane systems, it is likely that the use of compressed air for ground based, supercooled fog clearing will increase. Its use as an airborne system will be severely limited by the need to penetrate the cloud to deliver the ice crystals where they are needed.

14.4.4 Endothermic Materials

Materials which absorb heat as they go into solution are capable of producing localized cooling. Urea, or carbamide (NH_2CONH_2), possesses the very large endothermic heat of solution of 60.5 calories/g in water. When a powder of urea is dusted into a supercooled cloud, ice crystals are produced at temperatures as warm as 0°C (Knollenberg, 1966), although Langer et al. (1963) considered urea to be only partially active at -18°C.

Nucleation of ice by urea appears to require both the local cooling caused by dissolution of a particle in a drop and the epitaxial nucleation of ice on the urea crystal in the region of maximum temperature depression. It is not clear from the discussion given by Knollenberg (1966) if the cooling effect can extend beyond the drop or if it is adequate to cause nucleation by other ice nuclei present in either the drop or the air.

Field tests of milled urea powder with a mean size of 25-30 μm diameter and 10^7 to 10^8 particles/g were carried out on three days in February 1965 in northern and central Wisconsin by Knollenberg (1966). Each of the three stratoform clouds seeded produced strong snow showers shortly after 5 to 10 pounds urea per mile was released into the upper 200 feet of the clouds at temperatures between -13°C and -16°C. The results were impressive enough to claim urea was as effective as dry ice seeding.

However, urea has not received much attention for seeding cold clouds. Instead, its hygroscopic nature has been put to use in dissipating warm fogs (Weinstein and Silverman, 1972) and in modifying warm clouds (Clark et al., 1972). Since urea is a natural component of the biosphere, its use in seeding programs should present no major ecological difficulties.

14.5 BIOGENIC ICE NUCLEANTS

Since ice nuclei were considered vital to precipitation formation in temperate latitudes, inorganic materials originating at the surface of the earth were thought to be the principal source for natural ice nuclei. However, Schnell and Vali (1976) have presented convincing arguments for a large biogenic component to natural ice nuclei. Several bacteria are known to nucleate ice at temperatures above -2°C (Maki et al., 1974; Lindow et al., 1975, 1976; Arny et al., 1976; Vali et al., 1976). Ice nuclei derived from bacterial decay of leaves (Schnell and Vali, 1972) and from ocean microorganisms (Schnell et al., 1975) have also been considered potential components to the natural ice nucleus population.

The bacteria active in ice nucleation which are found in association with leaf derived nuclei (Schnell and Vali, 1972) were identified as *Pseudomonas syringae* by Maki et al. (1974). Other bacteria have also been found to show activity as ice nuclei: *P. fluorescens* (Maki and Garvey, 1975), *P. coronafaciens* (Arny et al., 1976), and *Erwinia herbicola* (Lindow et al., 1976).

Leaf-derived and ocean-derived ice nuclei are too small to be whole bacteria (Schnell and Vali, 1972; Schnell et al., 1975). However, these nuclei do appear to be produced during bacterial decay processes in some leaf litter and marine systems. Certain treatments which render leaf-derived nuclei inactive, such as autoclaving, point to the biological origin and organic nature of these nuclei, but their precise composition is still not known.

The high onset temperature and high ice nucleation activity at temperatures above -5°C are strong incentives for developing biogenic systems for ice nucleus production. However, the optimum conditions for producing bacterial nucleants or nuclei derived from bacterial systems are not yet well defined. Furthermore, it is not known what effect the dispersal of bacterial suspensions or other biogenic nuclei into the air has on the ice nucleation activity of these materials. Most previous studies have used these agents as freezing nuclei and only recently has Schnell (1976) shown that bacteria may serve as nuclei in clouds.

As with some of the other seeding agents discussed previously, the dispersal and delivery of biological agents to target clouds is a major hurdle to be crossed before operational use can be considered. Unlike organic and inorganic crystalline ice nucleants, the dispersal of bacterial ice nucleants requires the resulting aerosol particles to contain at least one whole bacterium. No matter how effectively the dispersal system is capable of producing single bacterium aerosols, the mass efficiency can not exceed about 10^{11}/g of pure bacteria. Dispersal of a suspension of bacteria in a carrier fluid will reduce the mass efficiency to between 10^5 to 10^8/g suspension. No estimate of the mass efficiency of leaf or ocean derived nuclei can presently be made although it is likely to be greater than that for whole bacteria at temperatures less than about -10°C.

Biogenic ice nuclei are already a part of the ecosphere (Schnell and Vali, 1976), hence their use as cloud seeding agents would not introduce new materials to the ecosystem but rather alter the quantity of these materials now present. Leaf- and ocean-derived nuclei are apparently part of the biodegradation sequence and thus should not pose any environmental hazard. However, some of the bacteria, notably *P. syringae*, are pathogenic to certain host plants under favorable conditions.

The intentional introduction of additional viable, pathogenic, and ice nucleation-active bacteria for the purpose of cloud seeding well may be considered imprudent. But, before this source of ice nuclei is discarded, the normal natural distribution of these bacteria on various plants as well as in the air must be considered. If the pathogen is already wide spread and readily available in the ecosphere, the relatively small additions associated with cloud seeding are not likely to upset the natural balance of epiphytic bacterial populations.

So little is presently known about biogenic ice nuclei that an estimate of their cost effectiveness can not be made. It must be noted, however, that if biogenic nuclei do play a significant role in cloud processes, as Schnell and Vali (1976) speculate, then agricultural, forest, rangeland, and selected ocean areas have always been natural sources for these nuclei. An increased output by manipulation of these natural producers of biogenic ice nuclei may be a feasible way to accomplish large area, low intensity cloud seeding at relatively low cost.

14.6 CONCLUSIONS

Several characteristics of an ice nucleating agent enter into the overall evaluation of its performance as a seeding agent: (1) onset temperature for ice nucleation, (2) shape of its activity spectrum as an ice nucleus aerosol, (3) efficiency of dispersing the agent to yield ice crystals, (4) flexibility of place of release with respect to the cloud being seeded, (5) cost per ice crystal produced in the cloud, and (6) the ecological impact of the agent over both short and long duration seeding programs. Ecological impact being the focus of the present volume, the other characteristics are important chiefly for their role in determining how much of what agent may come to be dispersed in the environment by cloud seeders.

Most of the alternative materials discussed in this chapter show onset temperatures for ice nucleation that equal or surpass that for silver iodide. Chilling agents, with dry ice the most outstanding example, can create ice crystals in clouds (or even in air only saturated with respect to ice) at temperatures just slightly below 0°C. Although chilling agents show the highest onset temperatures, the ice crystal production capability of most seeding agents at temperatures below about -8°C reaches a plateau which is generally less than the unit mass activity of organic and inorganic crystalline agents, including silver iodide. However, this may be a desirable feature since in most seeding projects a specific concentration of crystals to yield optimum growth is the desired goal of the seeding effort and it may be easier to control the concentration of crystals produced by using chilling agents. A few of the crystalline organic agents retain a unit mass activity advantage over silver iodide above -12°C. At temperatures below -12°C, however, silver iodide remains the most active ice nucleant on a per mass basis presently available. New inorganic or organic ice nucleants with characteristics superior to those presently known are not likely to be found except among the relatively unexplored organometallics of which copper acetylacetonate is an example.

The ease of dispersing silver iodide aerosols from ground-based and aircraft mounted solutions burners or by flares attached to or dropped from aircraft over, under, or within the target clouds gives silver iodide a distinct operational advantage in most applications over other ice catalysts. Present generator technology for organic nucleants limits dispersal by aircraft to near cloud base or during cloud penetration. Aerosol generation methods for the TiO_2-based silver iodide particles and the CuI substituted crystal appear adaptable to ground level, subcloud, or cloud penetrative seeding. Liquid chilling agents and adiabatic expansion of compressed air may be advantageous for dispersing supercooled fogs but are constrained to be released at temperatures below 0°C and in locations where the air is above ice saturation and the crystals produced can enter the cloud before subliming. The fall speed and life time of pellets of dry ice, however, allow this chilling agent to also be dropped into clouds during on-top seeding.

All of the alternative materials, when purchased in large quantities, are priced per unit mass less than silver iodide. However, when converted to a cost per ice crystal produced basis, many of the alternatives become more expensive than silver iodide. Only at relatively high temperatures, -8°C or above, do organic or chilling agents offer a cost advantage relative to silver iodide. Inorganic agents in which less expensive materials replace part of the silver iodide and which show ice nucleation activity spectra nearly equivalent to silver iodide may retain their cost advantage at all temperatures.

Among ice nucleants used in cloud seeding, only dry ice and adiabatic cooling of compressed air can be given a clean ecological bill of health. So little is known about the long-term ecological impact of other seeding agents that it is impossible to unequivocally state that they offer any significant proven long range ecological advantage over silver iodide.

Crystalline organic nucleants do offer the prospect of biodegradation but the pathways and daughter products of the process are not well known.

None of the alternative seeding agents discussed in this chapter is likely to replace silver iodide in all cloud modification projects in which an increased ice crystal concentration is the desired goal. However, in certain situations the advantage of a higher onset temperature and greater ice nucleation activity at high temperatures may lead to gradually expanding use of organic ice nucleants. Also, the uniform ice crystal production by some of the chilling agents over a broad temperature range may favor the use of these materials over other agents in certain applications. As the science of weather modification identifies the optimal seeding function for various seeding opportunities some of the alternative agents discussed in this chapter may well be chosen over silver iodide as the seeding agent for particular operational and research tasks. The result of this study suggests that neither the use of alternative ice nucleants for cloud seeding nor the substitution of them for silver iodide is likely to have a significant or discernible impact on the environment.

Summary

Weather modification contributions to environmental silver levels are relatively minor, however, individuals have unique relationships with atmospheric processes which may increase their awareness of the release of such chemicals to the environment. Silver iodide use in these programs will undoubtedly continue, although other seeding agents, notably dry ice, may be used to a greater extent in the future. The following areas may experience environmental impacts due to accumulations of nucleating agents.

Atmospheric Transport Processes

The variable background concentrations of silver in air and precipitation make detection of seeding materials extremely difficult. In mountainous areas, localized movement and limited dispersion of material can occur, and extended area effects on atmospheric processes are also possible. The release of tracer materials to differentiate seeding-derived silver from natural and other anthropogenic sources is recommended.

Chemical Equilibria in Aquatic and Terrestrial Environments

In both of these environments, chloride, bromide, and iodide ions control the amounts of free silver ion present, subject to the effects of physical parameters and other anions and cations. Silver can also form many complexes which are either too soluble or unstable, except under specialized Eh and pH conditions. Crystalline silver [Ag(c)] and silver sulfides can form under reduced conditions, and both decrease silver bioavailability. A computerized chemical model has been described for aquatic and estuarine environments which allows the prediction of chemical species at equilibrium, beginning with specific chemicals and defined abiotic conditions. In terrestrial, aquatic, and estuarine environments, probably only a minor fraction of total silver is in the free ionic form. Silver concentration totals given in most environmental analyses thus have little value for predicting reactive species, unless information about the available chemical species and the abiotic environment are included.

Biochemical Effects

Silver ion is the major silver species which reacts on a biochemical level, primarily with sulfhydryl groups, although it also effects hydrogen bonding, trace metal competition, and aggregation and depolymerization of macromolecules, including DNA. At low levels (10^{-5} M) silver ion has also been linked to the stimulation of biochemical reactions. After silver species in the environment reach chemical and physical equilibrium, effects of silver imposition at low levels may be difficult to detect; and it is possible that stimulation as well as inhibition might occur.

Effects on Microorganisms

Bacterial and fungal spores appear to be more resistant to silver ion than are vegetative forms, and low silver levels can stimulate microbial activity. These effects depend upon the silver compound used, the particular microorganism, and on the experimental conditions. Also, oxygen can affect the antimicrobial characteristics of various forms of silver. Microorganisms can develop tolerance to silver, which makes predicting environmental impacts more difficult.

Silver iodide from seeding generator burn complexes may produce antimicrobial effects but they are not observed consistently, and only with concentrations far exceeding levels expected to accumulate in natural environments.

Soil Microbiological Processes

Silver added to snow or rainfall through weather modification processes accumulates in surface soils. Subsequent rainfall will not cause appreciable silver movement below the surface soil zone. Silver increases detectable in precipitation are extremely difficult to detect in surface soils because available methodology cannot differentiate between silver forms at these levels, and because of higher background silver concentrations in soils. In addition, abiotic and biotic cycling of this metal makes it difficult to establish precise silver levels in terrestrial environments. Available laboratory and field studies show that silver iodide added to soil may cause slight decreases in microbial activity under short-term conditions (1 to 2 months), and that excess halide may have a role in this process. In contrast, silver ion can immediately affect soil microbial activity. Silver iodide added to field soils may lead to decreased litter decomposition rates in surface soils, which may result in increased rather than decreased soil microbiological activity.

The silver levels at which these effects might first be observed are at least 100-fold higher than levels normally occurring in the surface zone of the soils studied.

Plants

Silver concentrations in plants vary greatly. Silver from silver iodide is taken up by plants to a much lesser extent than is silver ion. With increases in environmental silver levels, relatively less silver is translocated to aboveground plant portions. More subtle aspects of metal accumulation and accommodation in plants may justify additional monitoring.

Silver ion can affect plant growth, phenological development, flowering, pollen tube germination, etc., and both stimulation and inhibition have been observed. Silver ion can influence ethylene production by plants and microbes and thus influence plant growth processes. Such silver treatment has been used commercially for plant preservation. Few experiments on possible silver iodide effects on plants have been conducted under carefully controlled laboratory or field conditions. In areas immediately surrounding previously used seeding generator sites where silver accumulation in surface soils is markedly higher than background levels, no effects on plant growth or vigor have been observed, although silver transport into aboveground plant parts has been noted.

Silver accumulates primarily in the root zone, probably in the elemental form; and no significant changes in silver levels in the plant-soil zone have been detected which can be related to weather modification activities, except near ground-based generator sites. Even if silver is imposed on terrestrial systems, the phenomenon of metal tolerance may make it difficult to predict or measure long-term silver impacts.

Silver and other metals may influence the plant epiphytic microflora, which might be of value for monitoring possible silver accumulation from precipitation. The atmospheric release by plants of microparticles containing metals, including silver, may lead to secondary nucleation phenomena.

Aquatic Environments

Silver entering aquatic environments will tend to be complexed by organic matter, and sediments contain much higher silver levels than do their associated waters. The silver content of both fresh and marine waters varies greatly, ranging from below detection limits to nearly 50 μg/liter. Numerous plants and primary consumer organisms in aquatic environments accumulate silver and many other metals at levels higher than those found in associated waters, and silver uptake does not appear to be regulated in some aquatic organisms. There is no strong evidence for silver bioaccumulation in higher trophic levels of such ecosystems, even when secondary consumers and predators feed on primary consumers which have accumulated silver above levels present in the abiotic environment.

Most available information about the toxicological effects of silver on aquatic organisms has been derived using silver ion, and these results are difficult to relate to assessments of silver iodide impacts. As organic matter and clay adsorb and inactivate silver ion, the ionic silver levels in these environments are expected to be quite low, and it will be difficult to show effects of silver iodide-derived nucleating agents on aquatic organisms under field conditions.

Effects on Sports Fish

In long-term (13 month) experiments with eyed rainbow trout eggs, silver iodide had a no effect concentration of between 0.03 and 0.06 μg/liter, within the limits of its own solubility. As embryonic acclimation to silver can occur, various no effect concentrations are possible. Considering the insolubility and low dissociation characteristics of silver iodide, this compound does not appear to be a threat to fish. Studies of a high mountain lake that had been impacted by weather modification programs support these laboratory conclusions.

Sewage Treatment Processes

Sewage treatment tends to concentrate heavy metals, increasing the possibility of microorganisms essential to these processes being affected. In spite of this, photographic effluents containing silver thiosulfate have not been shown to hinder sewage treatment processes.

Silver iodide from seeding agent burn complexes can inhibit methanogenesis, but only at high (500 μg/ml) levels. Lower levels of silver ion (32 to 64 μg/ml) can stimulate methanogenesis under defined cultural conditions.

The environmental impact of sludge disposal probably requires further investigation, since its use as a fertilizer may result in returning silver, and perhaps more importantly, other metals to the food web.

Humans

The effects on humans of ingestion or topical contact with silver iodide used in cloud seeding can be considered negligible. Decade-long observations of cases (unrelated to cloud seeding) of ingestion of large silver doses revealed no physiological effects, although some led to argyria, which is of cosmetic, but no physiological concern. In addition, surveys of seeding generator operators who have had long-term intensive contact with silver iodide reveal that they have not experienced medical difficulties.

Other Higher Terrestrial Organisms

Information is available on possible effects of silver iodide and burn agent complexes on rumen and caecum functions in animals. Because of the short retention times and the complexing ions present in the rumen, no effects from nucleating agents ingested with plant materials are expected.

The susceptability of a wide range of terrestrial organisms to silver ion has been investigated, but it is difficult to predict potential silver iodide effects based on this information. However, with the low levels of free ionic silver released from halide complexes and the reaction of silver ion with a wide range of organic and inorganic materials in natural environments, the effects of silver derived from weather modification activities are expected to be extremely difficult to detect or monitor, should they occur.

Nonsilver Iodide Nucleating Agents

A wide range of organic and inorganic compounds have been used as nucleating agents, in addition to silver iodide. All of these alternative materials, when purchased in large quantities, have lower per unit costs. However, on a cost per ice crystal-produced basis, many of the alternatives are more expensive than silver iodide. Only at relatively high temperatures, -8°C or above, do organic or chilling agents offer a cost advantage over silver iodide. Inorganic agents in which less expensive materials replace part of the silver iodide, and which show ice nucleation activity spectra nearly equivalent to silver iodide, may retain their cost advantage at all temperatures.

So little is known about the long-term ecological impact of other seeding agents that it is impossible to unequivocally state that they offer any significant advantage over silver iodide. Crystalline organic nucleants do offer the prospect of biodegradation, but the pathways and daughter products of the process are not well known.

Dry ice has many advantages under specific conditions, and is increasingly being used at the present time. The fall speed and life time of dry ice pellets allow its use for on-top seeding. Dry ice appears to be acceptable from an environmental standpoint, assuming that no toxic amine byproducts are present in the commercial seeding materials.

Overall Assessment

The major environmental concerns about nucleating agents (effects on plant growth, game animals, and fish, etc.) appear to represent negligible environmental hazards. The more subtle potential effects of silver-based nucleating agents, such as their possible ability to potentiate the movement or effects of other materials of environmental concern, or to influence the activity of microorganisms in soils and aquatic environments after being bioconcentrated by plants, warrant continued research and monitoring. Effects, if they occur, are not expected to involve unacceptable risks. The long-term use of silver iodide and the confidence which the weather modification profession has in delivery systems and in the efficacy of this material, make it unlikely that other agents, with the possible exception of dry ice, will be used on a large scale, unless there are improvements in delivery systems and major changes in the economics of silver availability.

Research Recommendations

Additional research or documentation on environmental impacts of nucleating agents used in weather modification programs may be required in the following specific areas:

- A continuing review should be made of literature concerned with nucleating agents, their use in weather modification programs and their potential impacts. This information should be made available perhaps on a bi-yearly basis.

- The development of new strategies for monitoring disposition, accumulation, and potential biological effects of seeding agents will be important, and these efforts should be carried out in conjunction with ongoing field programs. The monitoring of metal concentrations in surface mud microlayers, of possible accumulation effects on the plant leaf microbial flora, and the identification of more specific bioconcentration indicator plants, or sites on plants (the root zone) could be considered. In addition, measuring silver accumulation in tree rings may be possible under well-controlled conditions, and warrants further study.

- Continued improvement of analytical methodology appears to be needed. In large part, this technology will be available from other disciplines, however, a strong research effort will be required to apply these techniques to the analysis of silver levels and forms in soils and waters at the levels of resolution which are required. Research on the chemistry and biology of halide complexes (especially on their transformation and incorporation into general environmental metal pools) should be supported.

- A low-key but continuing effort should be made to evaluate the effects of nucleating agents on the impacts (positive or negative) of other materials released to the environment. Examples include other metals, herbicides, pesticides, and known carcinogens.

- If new seeding agents are proposed, a careful review of their possible ecological impacts should be made, based initially on information in the literature. Only when candidate agents have a strong likelihood of being used on an operational basis should specific research be supported. Agents being considered for widespread use should have markedly improve operational characteristics for demonstably less environmental risks.

- To aid in predicting long-term, low level effects of seeding agents in ecosystems, treatment plots and reference vegetation areas should be established in or adjacent to major impact zones. These should be monitored on perhaps 5 to 10 year intervals during the course of weather modification activities and also after completion of such programs. Ideally, these should be multipurpose sites where other environmental measurements are being made, which will increase the historical value of such information.

- The accumulation of data on weather modification activities, including areas seeded, and on the types and amounts of agents used, should be continued and expanded. This will provide information on trends which will aid in establishing research priorities and in predicting potential areas of public concern.

- The technology for tracing the fate of nucleating agents in seeded storm needs to be improved. Tracer compounds which normally occur in low concentrations in the impact ecosystems can be used, however, additional research is needed to improve the sensitivity of this technique.

- Basic studies should be carried out on the ability of plants, aquatic and terrestrial fauna, and microorganisms to accommodate or become tolerant to the presence of silver. In essence, biological systems can respond to and counteract the effects of added silver, resulting in changed responses with further silver additions--an extension of Weber's Law. This information will be of value in developing dose-response relationships in environmental impact statements.

- A number of key experiments should be conducted in each discipline to compare results derived from chemically pure silver iodide, silver iodide derived from AgI-NaI, and AgI-NH_4I burn complexes. With the continuing shift in seed solution mixtures to the use of the NH_4I complexing agent, some differences in biological responses might occur, although these are not expected to be of major concern.

- Efforts should be made to develop cooperative or complementary literature survey and research programs, involving the weather modification profession (public and private), the National Association of Photographic Manufacturers, the Silver Users Association, and the Federal funding agencies, as examples. This would assure maximum utilization and cross comparability of data on possible impacts of nucleating agents and related materials in the environment.

References

Adams, F. 1971. Ionic concentrations and activities in soil solutions. Soil Sci. Soc. Am. Proc. 35:420-426.

Addicks, L. 1940. Silver in industry. Reinhold Publishing Corp., New York. 636 p.

Adrian, W. J. 1971. A new wet digestion method for biological materials utilizing pressure. At. Absorpt. Newsl. 10(4):96.

Albright, L. J., J. W. Wentworth, and E. M. Wilson. 1972. Technique for measuring metallic salt effects upon the indigenous heterotrophic microflora of a natural water. Water Res. 6:1589-1596.

Albright, L. J., and E. M. Wilson. 1974. Sublethal effects of several metallic salts-organic compounds combinations upon the heterotrophic microflora of a natural water. Water Res. 8:101-105.

Allee, P. 1974. Ice nuclei concentrations measured by a benchmark network. Proc. 4th Conf. Weather Modification. Am. Meteorol. Soc., Fort Lauderdale, Fla. p. 242-245.

Allee, P., R. Sax, and E. Delgado. 1976. Concentration of Aitken, cloud droplet condensation and ice nuclei over selected oceanic and continental sites. Proc. Int. Conf. Cloud Physics. Am. Meteorol. Soc., Boulder, Colo. p. 2-8.

Anderson, B. G. 1948. The apparent thresholds of toxicity to *Daphnia magna* for chlorides of various metals when added to Lake Erie water. Trans. Am. Fish. Soc. 78:96-113.

Anderson, B. J., E. A. Jenne, and T. T. Chao. 1973. The sorption of silver by poorly crystallized manganese oxides. Geochim. Cosmochim. Acta 37:611-622.

Anderson, K. P., E. A. Butler, and E. M. Woolley. 1971. The complex solubility of silver chloride in methanol-water, acetone-water, and dioxane-water mixtures. J. Phys. Chem. 75:93-97.

Andrew, R. W. 1976. Toxicity relationships to copper forms in natural waters, chapter 6. *In* R. W. Andrew, P. V. Hodson, and D. E. Konasewich (eds.) Toxicity to biota of metal forms in natural waters. Proc. workshop held in Duluth, Minn., Oct. 7-8, 1975. Int. Joint Comm. Great Lakes Research Advisory Board, Windsor, Ontario. 329 p.

Andro, M., and R. Serpolay. 1970. Comparison des rendements en germes de glace obtenus par pulverisation de propane et d'antrydride carbonique dans un brouillard surfondu. J. Rech. Atmos. 4(2):93-100.

Angel, A., and K. J. Rogers. 1972. Analysis of the convulsant activity of substituted benzenes in the mouse. Toxicol. Appl. Pharmacol. 21(2):214-229.

Anghileri, L. J. 1969. In vivo breakdown of insoluble halides. Acta Isot. 9(4):347-356.

Annear, D. I., and B. J. Mee. 1976. Paradoxical response of bacteria to silver nitrate. Microbios Lett. 2(6):79-81.

Anonymous. 1962. Public health service drinking water standards. U.S. Public Health Serv. Publ. No. 956, Washington, D.C.

Anonymous. 1974. Silver in space-age medical applications. Silver Inst. Lett. 4(2):3.

Antikainen, P. J., and D. Dryssen. 1960. Studies on the hydrolysis of metal ions. Acta Chem. Scand. 14:86-101.

Antonovics, J., A. D. Bradshaw, and R. G. Turner. 1971. Heavy metal tolerance in plants. Adv. Ecol. Res. 7:1-85.

Appleman, H. S. 1968. The operational dissipation of supercooled fog. Proc. Int. Conf. Cloud Physics. Am. Meteorol. Soc., Toronto. p. 708-712.

Armstrong, J., N. Fukuta, and A. Gorove. 1974. Special organics field testing project. Final report to Div. Weather Modification. South Dakota Dep. Natural Resource Develop.

Arny, D. C., S. E. Lindow, and C. D. Upper. 1976. Frost sensitivity of *Zea mays* increased by application of *Pseudomonas syringae*. Nature (Lond.) 262:282-284.

Ashida, J., and H. Nakamura. 1959. Role of sulfur metabolism in copper resistance of yeasts. Plant Cell Physiol. 1:71-79.

Ashworth, L. J., and J. V. Amin. 1964. A mechanism for mercury tolerance in fungi. Phytopathology 54:1459-1463.

Avakyan, Z. A. 1967. Comparative toxicity of heavy metals for certain microorganisms. Mikrobiologiya 36(3):446-450.

Baptist, J. N., P. R. Steyermark, and P. L. Veltman. 1966. Microbiocides in hydrocarbon fuel compositions. U.S. Patent 3,226,210, Dec. 28, 1965. 4 p. (Chem. Abstr. 64:7953a).

Bard, C. C., J. J. Murphy, D. L. Stone, and C. J. Terhaar. 1976. Silver in photoprocessing effluents. J. Water Pollut. Control Fed. 48:389-394.

Barkov, G. D., and L. I. El'piner. 1968. The need for limiting the silver content of drinking water. Hyg. Sanit. 33:16-21.

Barthakur, N., and J. Maybank. 1963. Anomalous behavior of some amino acids as ice nucleators. Nature (Lond.) 200:866-868.

Barthakur, N., and J. Maybank. 1965. The growth of ice crystals on amino acid substrates. J. Rech. Atmos. 2:475.

Bashkirova, G. M., and P. N. Kariskov. 1957. Investigation of certain substances as crystallization reagents of supercooled fog. Trudy Glavnoi Geofiz. Obs. 72:118-126.

Battan, L. J. 1977. Weather modification in the Soviet Union-1976. Bull. Am. Meteorol. Soc. 58:4-19.

Beauford, W., J. Barber, and A. R. Barringer. 1977. Release of particles containing heavy metals from vegetation into the atmosphere. Science 195:571-573.

Bella, D. A. 1975. Tidal flats in estuarine water quality analysis. U.S. Environmental Protection Agency EPA-660/3-75-025. 186 p.

Bender, J. A. 1975. Trace metal levels in beach dipterans and amphipods. Bull. Environ. Contam. Toxicol. 14(2):187-192.

Berezin, I. V., Y. V. Savin, and K. Martinek. 1971. Two-component inhibition as a method for studying of enzyme active center flexibility. FEBS (Fed. Eur. Biochem. Soc.) Lett. 14(3):178-180.

Berne, E., and I. Leden. 1953. The formation of complexes between silver and bromide ions in aqueous solutions. Z. Naturforsch. 8a:719-726.

Bertine, K. K., and E. D. Goldberg. 1972. Trace elements in clams, mussels, and shrimp. Limnol. Oceanogr. 17:877-884.

Bertuzzi, A. 1936. Ric. Sci. Progr. teen. Econ. naz. 2(7):II325. (Cited in Zimmerman, 1952).

Beveridge, T. J., and R. G. Murray. 1976. Uptake and retention of metals by cell walls of *Bacillus subtilis*. J. Bacteriol. 127(3):1502-1518.

Beyer, E. M., Jr. 1976a. A potent inhibitor of ethylene action in plants. Plant Physiol. 58:268-271.

Beyer, E. M., Jr. 1976b. Silver ion: A potent antiethylene agent in cucumber and tomato. HortScience 11(3):195-196.

Biedermann, G., and L. G. Sillens. 1960. A critical survey of the solubility equilibrium of Ag_2O. Acta Chem. Scand. 14:717-725.

Bigg, E. 1973. Ice nucleus measurement in remote areas. J. Atmos. Sci. 30:1153-1157.

Black, W. A. P., and R. L. Mitchell. 1952. Trace elements in the common brown algae and seawater. J. Mar. Biol. Assoc. 30:575-584.

Blumberg, H., and T. N. Cary. 1934. Argyremia. J. Am. Med. Assoc. 103:1521-1524.

Bogen, J. 1974. Trace elements in precipitation and cloudwater in the area of Heidelberg measured by instrumental neutron activation analysis. Atmos. Environ. 8:835-844.

Bollay, E. 1965. Neutron activation analysis of silver content in precipitation samples. E. Bollay Associates. Final report to U.S. Dep. Interior Bur. Rec. Contract No. 14-06-D-5573. 45 p.

Bowen, H. J. M. 1966. Trace elements in biochemistry. Academic Press, Inc., New York. 241 p.

Boyle, R. W. 1968. The geochemistry of silver and its deposits. Geol. Surv. Can. Dep. Energy, Mines and Resour. Bull. No. 160, Queen's Printer, Ottawa. 264 p.

Bradford, G. R. 1971. Trace elements in the water resources of California. Hilgardia 41(3):45-53.

Bradford, G. R., F. L. Bair, and V. Hunsaker. 1968. Trace and major element content of 170 High Sierra lakes in California. Limnol. Oceanogr. 17(14):526-530.

Bragg, P. D., and D. J. Rainnie. 1974. The effect of silver ions on the respiratory chain of *E. coli*. Can. J. Microbiol. 20:883-889.

Braham, R. R., Jr. 1963. Phloroglucinol seeding of undercooled clouds. J. Atmos. Sci. 20:563-568.

Brier, G., and L. O. Grant. 1973. An evaluation of the extended area effects problem as related to the National Hail Research Experiment. Nat. Center Atmos. Res., Boulder, Colo. Interim Rep. No. 32-72. 66 p.

Brooks, R. R., and M. G. Rumsby. 1965. The biogeochemistry of trace element uptake by some New Zealand mollusks. Limnol. Oceanogr. 10:521-527.

Brown, R. W., and R. A. Anderson. 1968. The bactericidal effect of silver ion on *Pseudomonas aeruginosa*. J. Pharm. Pharmacol. 20(Suppl.):215-235.

Brown, W. V., H. Mollenhauer, and C. Johnson. 1962. An electron microscope study of silver nitrate reduction in leaf cells. Am. J. Bot. 49(1):57-63.

Bruland, K. W., K. K. Bertine, M. Koide, and E. D. Goldberg. 1974. History of metal pollution in Southern California coastal zone. Environ. Sci. Technol. 8:425-432.

Brun, A., and U. Brunk. 1968. Histochemical indications for lysosomal localization of heavy metals in normal rat brain and liver. J. Histochem. Cytochem. 18(11):820-827.

Bruni, A. 1938. Effetti dell'irradiazione dell'argento con raggi ultravioletti sulle proprieta oligodinamiche verso i batteri. Ann. d'igiene 48:733-742.

Bruni, A. 1942. Oligodynamic action of silver on alcoholic fermentations. Ricerca Sci. 13:322-327.

Bryan, G. W. 1974. Adaptation of an estuarine polychaete to sediments containing high concentrations of heavy metals, p. 123-135. *In* F. J. Vernberg and W. B. Vernberg (eds.) Pollution and physiology of marine organisms. Academic Press, Inc., New York.

Bryan, G. W. 1975. Some aspects of heavy metal tolerance in aquatic organisms, p. 7-34. *In* A. P. M. Lockwood (ed.) Effects of pollutants on aquatic organisms. Cambridge Univ. Press, Cambridge.

Buckley, W. R. 1963. Localized argyria. I. Chemical nature of the silver containing particles. Arch. Dermatol. 88:531-539.

Buckley, W. R., C. F. Oster, and D. W. Fassett. 1965. Localized argyria. II. Chemical nature of the silver containing particles. Arch. Dermatol. 92:697-705.

Buckley, W. R., and C. J. Terhaar. 1973. The skin as an excretory organ in argyria. Trans. St. John's Hospital Dermatol. Soc. 59:39-44.

Bunyan, J., A. T. Diplock, M. A. Cawthorne, and J. Green. 1968. Vitamin E and stress: Nutritional effects of dietary stress with silver in vitamin E-deficient chicks and rats. Br. J. Nutr. 22:165-182 (Paper No. 8).

Burris, N. 1977. *Chromobacterium:* Respiratory physiology and silver-stimulated violacein synthesis. M.S. thesis. Colorado State Univ., Fort Collins. 118 p.

Buschke, A., F. Jacobson, and E. Klopstock. 1925. The nature of oligodynamic antibacterial effect of metals. Dtsch. Med. Wochenschr. 51:595-598.

Calabrese, A., R. S. Collier, D. A. Nelson, and J. R. MacInnes. 1973. The toxicity of heavy metals to embryos of the American oyster *Crassostrea virginica*. Mar. Biol. 18:162-166.

Calabrese, A., J. R. MacInnes, D. A. Nelson, and J. E. Miller. 1977a. Survival and growth of bivalve larvae under heavy-metal stress. Mar. Biol. 41:179-184.

Calabrese, A., and D. A. Nelson. 1974. Inhibition of embryonic development of the hard clam, *Mercenaria mercenaria*, by heavy metals. Bull. Environ. Contam. Toxicol. 11(1):92-97.

Calabrese, A., F. P. Thurberg, and E. Gould. 1977b. Effects of cadmium, mercury, and silver on marine animals. Mar. Fish. Rev. 39:5-11.

Cameron, D. R. 1973. Modeling silver transport in the soil. Ph.D. thesis. Colorado State Univ., Fort Collins. 116 p.

Carroll, G. L. 1970. Pollution potential of propane used as a supercooled fog dispersal agent. Tech. Note. U.S. Army Cold Regions Research and Eng. Lab., Hanover, N.H. 7 p.

Castignetti, D. 1977. Silver ion and silver iodide burn complex effects on methanogenesis. M.S. thesis. Colorado State Univ., Fort Collins. 74 p.

Chambers, C. W., C. M. Proctor, and P. W. Kabler. 1962. Bactericidal effect of low concentrations of silver. J. Am. Water Works Assoc. 54:208-216.

Chappell, C. 1970. Modification of cold orographic clouds. Ph.D. dissertation. Dep. Atmos. Sci., Colorado State Univ., Fort Collins. 196 p.

Chappell, J. B., and G. D. Greville. 1954. Effect of silver ions on mitochondrial adenosine triphosphatase. Nature (Lond.) 174:930.

Charak, M. T. 1975. Weather modification activity reports: Calender year 1974. Nat. Ocean. Atmos. Admin. Office Environ. Monitoring and Prediction, Rockville, Md.

Charak, M. T., and M. DiGiulian. 1974. Weather modification activity reports: November 1, 1972 to December 31, 1973. Nat. Ocean. Atmos. Admin. Office Environ. Monitoring and Prediction, Rockville, Md.

Chateau, H., M. Durante, and B. Hervier. 1956a. Study of aqueous solutions of sodium sulfite. Sci. et Ind. Phot. 27:81-90.

Chateau, H., M. Durante, and B. Hervier. 1956b. Nature and stability of silver sulfite complexes. II. Sodium bisulfite solutions. Sci. et Ind. Phot. 27:257-262.

Chen, R. F. 1975. Effect of silver and mercury ions on the luminescence of tryptophan and proteins. Fluoresc. News 9(2-3):9-12.

Chessin, H., and B. Vonnegut. 1971. Lattice spacings of pseudobinary solid solutions of silver bromide and silver iodide. J. Am. Chem. Soc. 93:4964.

Christensen, H. E., and T. T. Luginbyhl (eds.). 1975. Registry of toxic effects of chemical substances. Nat. Inst. Occupational Safety and Health, Publ. Health Serv., Rockville, Md. 1296 p.

Church, B. D., L. Griffin, and J. W. King. 1975. The biodegradability and ecological effects of organic weather modification agents. Final Rep. Denver Research Inst., Univ. Denver. 92 p.

Clark, R. S., P. St-Amand, and T. L. Wright. 1972. Modification of warm convective clouds by hygroscopic and hydrophilic materials. Proc. 3rd Conf. Weather Modification. Am. Meteorol Soc., Rapid City, S.D. p. 179-181.

Cliver, D. O., W. K. Foell, and J. M. Goepfert. 1971. Biocidal effects of silver. Final Tech. Rep. Contract NAS 9-9300, Univ. Wisconsin, Madison. 64 p.

Cloke, P. L. 1963. The geological role of polysulfides. Part I. The distribution of ionic species in aqueous sodium polysulfide solution. Geochim. Cosmochim. Acta 27:1265-1299.

CODATA. 1971. CODATA task group on key values for thermodynamics. J. Chem. Thermo. 4:331-333.

CODATA. 1976. Report of the CODATA task group on key values of thermodynamics. CODATA Bull. No. 17.

Coleman, R. L., and J. E. Cearly. 1974. Silver toxicity and accumulation in largemouth bass and bluegill. Bull. Environ. Contam. Toxicol. 12:53-61.

Connick, R. E., and A. D. Paul. 1961. The fluoride complexes of silver and stannous ions in aqueous solutions. J. Phys. Chem. 65:1215-1220.

Cooper, C. F., and W. C. Jolly. 1969. Ecological effects of weather modification: A problem analysis. School of Natural Resources, Univ. Michigan. 160 p.

Cooper, C. F., and W. C. Jolly. 1970. Ecological effects of silver iodide and other weather modification agents: A review. Water Resour. Res. 6(1):88-98.

Corelli, J. P. 1971. A laboratory study to determine the disposition of silver iodide in selected soils. Prof. Pap. Dep. Watershed Sci., Colorado State Univ., Fort Collins. 32 p.

Corrin, M. L., and R. Barnes. 1976. *In situ* silver iodide-coated titanium dioxide as an ice nucleant. J. Appl. Meteorol. 15:413-414.

Cowgill, U. M. 1973. Biogeochemical cycles for the chemical elements in *Nymphaea odorata* Ait. and the aphid *Rhopalosiphum nymphaeae* (L.) living in Linsley Pond. Sci. Total Environ. 2:259-303.

Cowgill, U. M., and C. W. Burns. 1975. Differences in chemical composition between two species of *Daphnia* and some freshwater algae cultured in the laboratory. Limnol. Oceanogr. 20:1005-1011.

Cuta, F., E. Berganek, and J. Pisecky. 1957. Determination of the thermodynamic second dissociation constant of sulfurous acid from potentiometric and spectrophotometric measurements. Chem. Listy 51:1614-1617.

Danielli, J. F. 1944. The biological action of ions and the concentration of ions at surfaces. J. Exp. Biol. 20:167-176.

Danielli, J. F., and J. T. Davies. 1951. Reactions at interfaces in relation to biological problems. Adv. Enzymol. 11:35-89.

Daune, M., C. A. Dekker, and H. K. Schachman. 1966. Complexes of silver ion with natural and synthetic polynucleotides. Biopolymers 4:51-76.

Davies, C. W. 1962. Ion association. Butterworths, Inc., Washington, D.C. 190 p.

Davies, P. H. 1975. Trace metal analysis in water, using organic extraction with APDC-MIBK. Colo. Div. Wildl. Fish. Inform. Leafl. No. 31. 4 p.

Davies, P. H. 1976. A method for determining the long-term toxicity of insoluble metal compounds demonstrated by silver iodide in natural water, p. 118-120. *In* R. W. Andrew, P. V. Hodson, and D. E. Konasewich (eds.) Toxicity to biota of metal forms in natural water. Proc. workshop held in Duluth, Minn., Oct. 7-8, 1975. Int. Joint Comm. Great Lakes Research Advisory Board, Windsor, Ontario. 329 p.

Davies, P. H., J. P. Goettl, Jr., and J. R. Sinley. 1978. Toxicity of silver to rainbow trout. Water Res. (in press).

Davies, P. H., J. P. Goettl, Jr., J. R. Sinley, and N. F. Smith. 1976. Acute and chronic toxicity of lead to rainbow trout (*Salmo gairdneri*) in hard and soft water. Water Res. 10:199-206.

Davis, G. K. 1972. Competition among mineral elements relating to absorption by animals. Ann. N.Y. Acad. Sci. 199:62-69.

Demuynck, M., K. Rahn, M. Janssens, and R. Dams. 1976. Chemical analysis of airborne particulate matter during a period of unusually high pollution. Atmos. Environ. 10:21-26.

Dessens, J., and J. P. Lacaux. 1972. Ground seeding for hail prevention in south-western France: Possible overstepping of an economic efficiency level from 1963. Proc. 3rd Conf. Weather Modification. Am. Meteorol. Soc., Rapid City, S.D. p. 268-271.

Dick, R. 1975. Sludge treatment, utilization, and disposal. J. Water Pollut. Control Fed. 47:1306-1323.

Dickinson, C. H., B. Austin, and M. Goodfellow. 1975. Quantitative and qualitative studies of phylloplane bacteria from *Lolium perenne*. J. Gen. Microbiol. 91:157-166.

Dickinson, C. H., and T. F. Preece (eds.). 1976. Microbiology of aerial plant surfaces. Academic Press, Inc., New York. 669 p.

Diem, K. (ed.). 1962. Documenta geigy: Scientific tables. Geigy Pharmaceuticals, Ardsley, N.Y. 778 p.

Dinman, B. D. 1972. "Non-concept" of "no-threshold": Chemicals in the environment. Science 175:495-497.

Diplock, A. T., J. Green, J. Bunyan, D. McHale, and I. R. Muthy. 1967. Vitamin E and stress. III. The metabolism of D-α-tocopherol in the rat under dietary stress with silver. Brit. J. Nutr. 21(1):115-125.

Dirkse, T. P. 1962. The AgO-Ag_2O electrode in alkaline solution. J. Electrochem. Soc. 109:173-177.

Dirkse, T. P., and W. M. Graven. 1959. The stability and solubility of AgO in alkaline solutions. J. Electrochem. Soc. 106:284-287.

Dobryanskii, V. M. 1974. Data for hygenic standardization of metaldehyde in the air of the working zone. Gig. Tr. Prof. Zabol 2:55-56.

Doerr, R. 1920a. Zur oligodynamie des Silbers. Biochem. Z. 106:110-133.

Doerr, R. 1920b. Zur oligodynamie des Silbers. II. Mitteilung. Biochem. Z. 107:207-218.

Doerr, R. 1921. Zur oligodynamie des Silbers. III. Mitteilung. Biochem. Z. 113:58-69.

Doerr, R., and W. Berger. 1922. Zur oligodynamie des Silbers. IV. Mitteilung. Biochem. Z. 131:351-361.

Douderoff, D., and M. Katz. 1953. Critical review of literature on the toxicity of industrial wastes and their components to fish. Sewage Ind. Wastes 25:802-839.

Douglas, W. J. 1968. Toxic properties of materials used in weather modification. Proc. 1st Conf. Weather Modification. Am. Meteorol. Soc., Albany, N.Y. p. 351-360.

Douglas, W. J. 1970. Silver iodide generators and public health. ASCE Irrig. Drainage Div. J. 96(3):273-280.

Duce, R. A., J. G. Quinn, C. E. Olney, S. R. Piotrowicz, B. J. Ray, and T. L. Wade. 1972. Enrichment of heavy metals and organic compounds in the surface microlayer of Narragansett Bay, Rhode Island. Science 176:161-163.

Durbin, P. W. 1960. Metabolic characteristics within a chemical family. Health Phys. 2:225-238.

Eddie, W. J., and T. R. Mee. 1963. The effect of dry ice pellet velocity on the generation of ice crystals. J. Appl. Meteorol. 2:260-265.

Eichhorn, G. L. 1975. Active sites of biological macromolecules and their interaction with heavy metals. Environ. Sci. Res. F (Ecol. Toxicol. Res.) 7:123-142.

Eichhorn, G. L., P. Clark, and E. Tarien. 1969. The interaction of metal ions with polynucleotides and related compounds. J. Biol. Chem. 244:937-942.

Elliott, R. D., R. W. Shaffer, A. Court, and J. F. Hannaford. 1976. Colorado River Basin pilot project. Comprehensive Evaluation Rep. No. ARI-76-1. U.S. Dep. Interior Bur. Rec., Denver, Colo. 641 p.

Ellis, E. L. 1962. Inorganic constituents of wood. For. Prod. J. 12:271-274.

Engelmann, R., D. Hagen, W. Haller, and R. Perkins. 1967. Washout coefficients for silver iodide. Presented at 47th Annu. Meeting Am. Meteorol. Soc. Tech. Rep. BNWL-SA-798. Battelle-Northwest Laboratory, Richland, Wash. p. 13.

Ernst, W. 1976. Physiological and biochemical aspects of metal tolerance. Soc. Exp. Biol., Seminar Series 1:115-133.

Evans, G. W., P. F. Majors, and W. E. Cornatzer. 1970. Mechanism for cadmium and zinc antagonism of copper metabolism. Biochem. Biophys. Res. Commun. 40(5):1142-1148.

Evans, L. F. 1966. Ice nucleation by amino acids. J. Atmos. Sci. 23:751-752.

Feitknecht, W., and P. Schindler. 1963. Solubility constants of metal hydroxide salts in aqueous solution. Pure Appl. Chem. 6:125-199.

Fins, L. 1972. Effects of cloudseeding materials on Engelmann spruce pollen germination. M.S. thesis. Colorado State Univ., Fort Collins. 77 p.

Fish, B. R. 1972. Electrical generation of natural aerosols from vegetation. Science 175:1239-1240.

Fishman, M. J., and M. R. Midgett. 1968. Extraction techniques for the determination of cobalt, nickel, and lead by atomic absorption, p. 230-235. *In* Trace inorganics in water. Advances in Chemistry Series No. 73, Am. Chem. Soc., Washington, D.C.

Fitzgerald, G. P. 1967. The algistatic properties of silver. Water Sewage Works 114(5):185-189.

Fitzgerald, G. P., and S. L. Faust. 1963. Bioassay for algicidal versus algistatic chemicals. Water Sewage Works 110:296-298.

Fitzgerald, G. P., and G. C. Gerloff, and F. Skogg. 1952. Studies on chemicals with selective toxicity to blue-green algae. Sewage Ind. Wastes 24:888-896.

Fjerdingstad, E., E. Fjerdingstad, and F. Popea. 1976. Analysis of heavy metals and bacteria in sediments from Danish lignite pits. Arch. Hydrobiol. 77(2):226-253.

Fleagle, R. G., J. A. Crutchfield, R. W. Johnson, and M. F. Abdo. 1974. Weather modification in the public interest. Am. Meteorol. Soc., Univ. Washington Press, Seattle. 88 p.

Fletcher, M. 1969. The physics of rain clouds. Cambridge Univ. Press, Cambridge. 390 p.

Fletcher, R. D. 1971. Operational application of fog modification. Preprints Int. Conf. Weather Modification. Am. Meteorol. Soc., Canberra, Australia. p. 255-258.

Forster, W. O., D. A. Wolfe, F. G. Lowman, and R. McClin. 1971. Trace element interactions between river water and sea water. Proc. 3rd Nat. Symp. Radioecology. CONF-710501-P2, Vol. II. Oak Ridge, Tenn. p. 807-815.

Freeman, R. A. 1975. Ecological kinetics of silver in an alpine ecosystem. Ph.D. thesis. Colorado State Univ., Fort Collins. 87 p.

Frutig, J. 1977. Sewage sludge yields up gold, silver, tin ... The Christian Science Monitor, 3 October.

Fujihara, M. P., T. R. Garland, R. E. Wildung, and H. Drucker. 1973. Response of microbiota to the presence of heavy metals in soil. Bacteriol. Proc. G37.

Fujita, S. 1971. Silver-palladium-gold alloys carcinogenicity and acid mucopolysaccharides in the induced tumors. Shika Igaku 34(6):918-932. (Chem. Abstr. 77:136075a).

Fukuta, N. 1963. Ice nucleation by metaldehyde. Nature (Lond.) 199:475-476.

Fukuta, N. 1965a. Activated ice nucleation by sprayed organics. J. Atmos. Sci. 22:207-211.

Fukuta, N. 1965b. Production of ice crystals in air by a pressure pack method. J. Appl. Meteorol. 4:454-456.

Fukuta, N. 1966. Experimental studies of organic ice nuclei. J. Atmos. Sci. 23:191-196.

Fukuta, N. 1967. An airborne generator of metaldehyde smoke. J. Appl. Meteorol. 6:948-951.

Fukuta, N. 1968. Some remarks on ice nucleation by metaldehyde. Proc. Int. Conf. Cloud Physics. Am. Meteorol. Soc., Toronto. p. 194-198.

Fukuta, N. 1972a. Metaldehyde seeding in summertime cumuli. J. de Rech. Atmos. 165-174.

Fukuta, N. 1972b. Advances in organic ice nuclei generator technology. J. de Rech. Atmos. 156-164.

Fukuta, N. 1975. A study of the mechanism of contact ice nucleation. J. Atmos. Sci. 32:1597-1603.

Fukuta, N., L. F. Evans, Y. H. Paik, W. A. Schmeling, L. A. Walter, and T. L. Wang. 1970. A study of the production and detection of artificial ice nuclei. Rep. No. 2547. Denver Research Inst., Univ. Denver.

Fukuta, N., K. J. Heffernan, W. J. Thompson, and C. T. Maher. 1966. Generation of metaldehyde smoke. J. Appl. Meteorol. 5:288-291.

Fukuta, N., and B. J. Mason. 1963. Epitaxial growth of ice on organic crystals. J. Phys. Chem. Solids 24:715-718.

Fukuta, N., and Y. H. Paik. 1974. Generating ice nuclei smoke particles for weather modification and apparatus therefore. U.S. Patent 3,835,059. U.S. Dep. Interior. 5 p.

Fukuta, N., M. N. Plooster, Y. H. Paik, C. F. Evans, and A. Gorove. 1973. Engineering, microphysical, and dynamical aspects of precipitation management. Denver Research Inst. Tech. Rep. No. 227783/8GA. 155 p.

Fukuta, N., W. A. Schmeling, and C. F. Evans. 1971. Experimental determination of ice nucleation by falling dry ice pellets. J. Appl. Meteorol. 10:1174-1179.

Furchner, J. E., C. R. Richmond, and G. A. Drake. 1968. Comparative metabolism of radionuclides in mammals. IV. Retention of silver-110m in the mouse, rat, monkey, and dog. Health Phys. 15(6):505-514.

Gaines, A. G., Jr., and M. E. Q. Pilson. 1972. Anoxic water in the Pettaquamscutt River. Limnol. Oceanogr. 17(1):42-49.

Gaivoronskii, I. I., L. I. Krasnovskaia, and A. D. Solov'ev. 1968. Artificial low cloud and fog dispersal. Proc. Int. Conf. Cloud Physics. Am. Meteorol. Soc., Toronto. p. 700-707.

Gammill, J. C., B. Wheeler, E. L. Carothers, and P. F. Hahn. 1950. Distribution of radioactive silver colloids in tissues of rodents following injection by various routes. Proc. Soc. Exp. Biol. Med. 74:691-695.

Gang, M. W., and D. Langmuir. 1974. Controls on heavy metals in surface and ground waters affected by coal mine drainage: Clarion River-Redbank Watershed, Pa. 5th Symp. on Coal Mine Drainage Research. Coal and the Environ. Tech. Conf., Louisville, Ky. p. 39-69.

Ganther, H. E., P. A. Wagner, M. L. Sunde, and W. G. Hoekstra. 1973. Protective effects of selenium against heavy metal toxicities, p. 247-252. *In* D. D. Hemphill (ed.) Trace substances in environmental health. VI. Proc. Univ. Missouri's 6th Annu. Conf. on Trace Substances in Environ. Health, Columbia, Mo.

Garrels, R. M., and C. R. Naeser. 1958. Equilibrium distribution of dissolved sulphur species in water at 25°C and 1 atmosphere total pressure. Geochim. Cosmochim. Acta 15:113-130.

Garrels, R. M., M. E. Thompson, and R. Siever. 1961. Solubility of carbonates in sea water; control by carbonate complexes. Am. J. Sci. 259:24-45.

Garten, V. A., and R. B. Head. 1965. A theoretical basis of ice nucleation by organic crystals. Nature (Lond.) 205:160-162.

Garvey, D. 1975. Testing of cloud seeding materials at the cloud simulation and aerosol laboratory, 1971-1973. J. Appl. Meteorol. 14:883-890.

Gatz, D. F. 1975. Background silver concentrations in Illinois precipitation and river water. J. Appl. Meteorol. 14(2):217-221.

Gerber, H. 1974. Size and activity of AgI particles. Proc. 4th Conf. Weather Modification. Am. Meteorol. Soc., Fort Lauderdale, Fla. p. 228-235.

Gerber, H., and P. Allee. 1972. Size distributions of artificial and natural ice nuclei. 3rd Conf. Weather Modification. Am. Meteorol. Soc., Rapid City, S.D. p. 24-29.

Gerber, H., P. Allee, U. Katz, C. Davis, and L. O. Grant. 1970. Some size distribution measurements of AgI nuclei with an aerosol spectrometer. J. Atmos. Sci. 27:1060-1067.

Gerdel, R. W. 1968. Note on the use of liquified propane for fog dispersal at the Medford-Jackson airport, Oregon. J. Appl. Meteorol. 7:1039-1040.

Giangiordano, R. A. 1976. Silver-temperature responses and acquired resistance of *Hyphomicrobium*. M.S. thesis. Colorado State Univ., Fort Collins. 137 p.

Gillen, K., R. Jensen, and N. Davidson. 1964. Binding of silver ion by adenine and substituted adenines. J. Am. Chem. Soc. 86:2792-2796.

Gingell, S. M., R. Campbell, and M. H. Martin. 1976. The effect of zinc, lead, and cadmium pollution on the leaf surface microflora. Environ. Pollut. 11:25-37.

Giordano, P. M., and J. J. Mortvedt. 1976. Nitrogen effects on mobility and plant uptake of heavy metals in sewage sludge applied to soil columns. J. Environ. Qual. 5:165-168.

Girvin, D. C., A. T. Hodgson, and M. H. Panietz. 1975. Assessment of trace metal and chlorinated hydrocarbon contamination in selected San Francisco Bay estuary shellfish. Energy Environ. Div., Lawrence Berkeley Lab., Univ. California, Berkeley, UCID-3778. 82 p.

Golterman, H. L., and R. S. Clymo (eds.). 1970. Chemical analysis of fresh waters. 2nd Rev. IBP Handbook No. 8. Blackwell Sci. Publ., Oxford. 166 p.

Golubovich, V. N. 1974. Toxic effect of ionic silver on various groups of microorganisms. Mikrobiologiya 43(5):922-924.

Golubovich, V. N., and I. L. Rabotnova. 1974. Kinetics of growth inhibition in *Candida utilis* by silver ions. Mikrobiologiya 43(6):115-117.

Gorbenko-Germanov, D. S., and N. K. Baskova. 1973. Oxidation of silver (+) ion by ozone and kinetics of decomposition of silver (+2) ion in aqueous perchloric acid solutions. Tezisy. Dokl. Vses. Soveshch. Khim. Neorg. Perekisnykh Soedin. p. 117-118.

Gordon, J. W. 1961. Thermodynamic functions of gaseous dioxides of carbon, nitrogen, sulfur, and chlorine, and of carbon disulfide and oxysulfides. J. Chem. Eng. Data 6:390-394.

Gorin, G., and C-C. Chin. 1965. Urease IV: Its reaction with n-ethylmaleimide and with silver ions. Biophys. Biochem. Acta 99:418-426.

Grasso, P., R. Abraham, R. Hendy, A. T. Diplock, L. Goldberg, and J. Green. 1969. The role of dietary silver in the production of liver necrosis in vitamin E deficient rats. Exp. Mol. Pathol. 11:186-199.

Gregor, L. V., and K. S. Pitzer. 1962. The silver-silver oxide electrode; the entropy of mercuric oxide. J. Am. Chem. Soc. 84:2671-2673.

Grimes, A. E. 1972. An annotated bibliography on the biological and physiological effects of silver iodide cloud seeding, 1962-1971. Nat. Ocean Atmos. Admin. Memo. EDS ESIC-2, U.S. Dep. Commerce. 21 p.

Grosch, M., and H. Georgii. 1976. Elemental composition of atmospheric freezing nuclei. Proc. Int. Conf. Cloud Physics. Am. Meteorol. Soc., Boulder, Colo. p. 673-675.

Guilloux, E., J. E. Courtois, and F. Percheron. 1968. Etude d'une trehalase purifiee extraite du hanneton (*Melolontha vulgaris*). Bull. Soc. Chim. Biol. 50(11):1915-1931.

Guyer, M. F., and F. E. Mohs. 1933. Rat carcinoma and injected colloidal platinum. Arch. Pathol. 15:796-817.

Habu, T. 1968. Histopathological effects of silver-palladium-gold alloy implantation on the oral submucous membranes and other organs. Shika Igaku 31(1):38-48. (Chem. Abstr. 69:95035b).

Hagashida, H., R. K. Semba, F. Niwa, and S. Kashiwamata. 1975. Mitochondrial malate dehydrogenase of bovine cerebrum: Characterization and mechanisms of inhibition by silver ions. J. Biochem. (Tokyo) 78:989-999.

Halcrow, W., D. W. Mackay, and I. Thornton. 1973. The distribution of trace metals and fauna in the Firth of Clyde in relation to the disposal of sewage sludge. J. Mar. Biol. 53:721-739.

Hamilton, E. I., and M. J. Minski. 1972. Comments upon the inorganic constituents present in DNA and RNA. Sci. Total Environ. 1:104-107.

Hamilton, E. I., M. J. Minski, and J. J. Cleary. 1973. The concentration and distribution of some stable elements in healthy human tissues from the United Kingdom. Sci. Total Environ. 1:341-374.

Harley, J. L., and D. H. Jennings. 1958. The effect of sugars on the respiratory response of beech mycorrhizas to salts. Proc. R. Soc. Lond., Ser. B, 148:403-418.

Harold, F. M. 1974. Chemiosmotic interpretation of active transport in bacteria. Ann. N.Y. Acad. Sci. 227:297-311.

Harrison, P. R., K. A. Rahn, R. Dams, J. A. Robbins, J. W. Winchester, S. S. Brar, and D. M. Nelson. 1971. Areawide trace metal concentrations measured by multielement neutron activation analysis: A one day study in northwest Indiana. J. Air Pollut. Control Assoc. 21(9):563-570.

Harry, H. W., and D. V. Aldrich. 1958. The ecology of *Australorbis glabratus* in Puerto Rico. Bull. W H O (World Health Organization) 18:819-832.

Head, R. B. 1961. Steroids as ice nucleators. Nature (Lond.) 191:1658-1659.

Head, R. B. 1962a. Ice nucleation by some cyclic compounds. J. Phys. Chem. Solids 23:1371-1378.

Head, R. B. 1962b. Ice nucleation by α-phenazine. Nature (Lond.) 196:736.

Helgeson, H. C. 1969. Thermodynamics of hydrothermal systems at elevated temperatures and pressures. Am. J. Sci. 267:729-804.

Hicks, J. R. 1967. Improving visibility near airports during periods of fog. J. Appl. Meteorol. 6:39-42.

Hicks, J. R., and G. Vali. 1973. Ice nucleation in clouds by liquified propane spray. J. Appl. Meteorol. 12:1025-1034.

Hill, C. B., B. Starcher, and G. Matrone. 1964. Mercury and silver interrelationships with copper. J. Nutr. 83(2):107-110.

Hodge, V., S. R. Johnson, and E. D. Goldberg. 1978. Influence of atmospherically transported aerosols on surface ocean water composition. Geochem. J. (in press).

Hoffman, L. E., and J. L. Hendrix. 1976. Inhibition of *Thiobacillus ferrooxidans* by soluble silver. Biotechnol. Bioeng. 18:1161-1165.

Hopkins, H. P., and C. A. Wulff. 1965. The solution thermochemistry of polyvalent electrolytes. II. Silver sulfate. J. Phys. Chem. 69:9-11.

Hopps, H. C., and H. C. Cannon (eds.). 1972. Geochemical environment in relation to health and disease. Ann. N.Y. Acad. Sci. 199:1-352.

Horak, D. L. 1969. Hatchery trout feeding levels. Colo. Div. Wildl. Fish. Inform. Leafl. No. 14. 4 p.

Horowitz, C. T., H. H. Schock, and L. A. Horowitz-Kisimova. 1974. The content of scandium, thorium, silver, and other trace elements in different plant species. Plant Soil 40:397-403.

Hosenfeld, M. 1938a,b. Die Oligodynamie in Theorie und Praxis: I and II. Chemiker Zeitung 62:3-5, 20-21.

Hosler, C. L. 1951. On the crystallization of supercooled clouds. J. Meteorol. 8:326.

Hough, D. W., and S. Shall. 1970. Inhibition of enzymatically active n-acetylhomocysteinyl ribonuclease by silver ions. FEBS (Fed. Eur. Biochem. Soc.) Lett. 8(5):243-246.

Howell, W. E. 1977. Environmental impacts of precipitation management: Results and inferences from project skywater. Bull. Am. Meteorol. Soc. 58:488-501.

Imai, K., T. Sugio, T. Tsuchida, and T. Tano. 1975. Effect of heavy metal ions on the growth and iron-oxidizing activity of *Thiobacillus ferrooxidans*. Agric. Biol. Chem. 39:1349-1354.

International Rice Research Institute. 1964. Annual report. Los Banos, Laguna, Philippines. 335 p.

Jaffre, T., R. R. Brooks, J. Lee, and R. D. Reeves. 1976. *Sebertia acuminata*: A hyperaccumulator of nickel from New Caledonia. Science 193:579-580.

Jasionowski, W. J., and E. T. Allen. 1974. Silver ion bactericide systems. ASME Special Publ. 74-ENAs-49.

Jenne, E. A., and S. N. Luoma. 1977. The forms of trace elements in soils, sediments, and associated waters: An overview of their determination and biological availability, p. 110-143. *In* R. E. Wildung and H. Drucker (eds.) Biological implications of metals in the environment. CONF-750929. NTIS, Springfield, Va.

Jensen, L. S. 1974. Interactions of silver and copper with selenium in chicks. Fed. Proc. 33:694.

Jensen, W. A., and L. G. Kavaljian. 1956. The cytochemical localization of ascorbic acid in root tip cells. J. Biophys. Biochem. Cytol. 2:87-92.

John, W., R. Kaifer, K. Rahn, and J. Wesolowski. 1973. Trace element concentrations in aerosols from the San Francisco Bay area. Atmos. Environ. 7:107-118.

Johnson, G. K. 1975. The standard enthalpy of formation of the aqueous iodide ion. Conf. Int. Thermodyn. Chim. 1:26-31.

Jones, A. M., and V. A. Bailey. 1974. The effect of silver from cloud seeding on cecal flora of rabbits. Water, Air, Soil Pollut. 3:353-363.

Jones, J. R. E. 1941. A further study of the relation between toxicity and solution pressure, with *Polycelis nigra* as test animal. J. Exp. Biol. 17:408-415.

Jonte, J. H., and D. S. Martin. 1952. The solubility of silver chloride and the formation of complexes in chloride solutions. J. Am. Chem. Soc. 74:2052-2054.

Junge, C. E. 1963. Air chemistry and radioactivity. Academic Press, Inc., New York. Int. Geophys. Ser. 4:382.

Just, J., and A. Szniolis. 1936. Germicidal properties of silver in water. J. Am. Water Works Assoc. 28:492-506.

Kaneshiro, C. M., C. A. Enns, M. G. Hahn, J. S. Peterson, and F. J. Reithel. 1975. Evidence for an active dimer of *E. coli* β-galactosidase. Biochem. J. 151(2):433-434.

Katz, V. 1960. Zur Eiskernbildungsfahigkeit von Kupferoxyden und Kupfersulfiden. Z. angew. Math. Phys. 11:237.

Kehoe, R. A., J. Cholak, and R. V. Story. 1940. A spectrochemical study of the normal ranges of concentration of certain trace metals in biological materials. J. Nutr. 19:579-592.

Kelley, K. K., and C. T. Anderson. 1935. Contributions to the data on the metallurgy. IV. Metal carbonates-correlations and applications of thermodynamic properties. U.S. Bur. Mines Bull. 384.

Kent, N. L. 1942. The occurrence of lead, tin, and silver in wheat and its milling products. J. Soc. Chem. Ind. 61:1-19.

Kent, N. L., and R. A. McCance. 1941. The absorption and excretion of minor elements by man. Biochem. J. 35:837.

Kharkar, K., K. Turekian, and K. Bertine. 1968. Stream supply of dissolved Ag, Mo, Sb, Se, Cr, Co, Rb, and Cs to the oceans. Geochim. Cosmochim. Acta 32:285-298.

Khovrychev, M. P., E. A. Andreeva, S. A. Lirova, V. N. Golubovich, V. A. Romazkova, V. A. Ryabckuk, and R. M. Fedorovich. 1976. Influence of H^+, OH^-, Cu^{+2}, and Ag^+ on the amino acid composition of cells of a chemostat culture of *Candida utilis*. Mikrobiologiya 45(3):437-439.

King, R., J. Fordyce, A. Antoine, H. Leibecki, H. Neustadter, and S. Sikik. 1976. Elemental composition of airborne particulates and source identification: An extensive one year survey. J. Air Pollut. Control Assoc. 26:1073-1078.

Kiss, I., and I. Peterfi. 1960. Inhibiting the activity of soil maltase. Pochvovedenie 8:84-86.

Klein, D. A., and R. A. Giangiordano. 1976. Evaluation of potential impacts of silver iodide nucleating agents on aerobic and anaerobic aquatic microbial processes. Proc. 2nd WMO Sci. Conf. Weather Modification, Boulder, Colo. p. 569-575.

Klein, D. A., and E. M. Molise. 1975. Ecological ramification of silver iodide nucleating agent accumulation in soil and aquatic environments. J. Appl. Meteorol. 14(5):673-680.

Klein, D. A., and R. A. Sokol. 1976. Evaluation of ecological impacts of silver iodide seeding agents, p. 125-138. *In* H. W. Steinhoff and J. D. Ives (eds.) Ecological impacts of snowpack augmentation in the San Juan Mountains, Colorado. Final Rep. San Juan Ecology Project, Colorado State Univ., Fort Collins.

Klein, D. A., and W. D. Striffler. 1976. Deposition and environmental impact of silver iodide in the National Hail Research Experiment. Operations Rep. No. 5, Colorado State Univ., Fort Collins. 61 p.

Klein, D. A., W. D. Striffler, and H. L. Teller. 1975. Disposition and environmental impact of silver iodide in the National Hail Research Experiment. Operations Rep. No. 4, Colorado State Univ., Fort Collins. 63 p.

Kleinkopf, M. D. 1960. Trace elements in Maine lake waters. Bull. Geol. Soc. Am. 71:1231-1242.

Kline, D. 1963. Evidence of geographical differences in ice nuclei concentrations. Mont. Weather Rev. 91:681-686.

Klusman, R. W., and K. W. Edwards. 1976. Toxic heavy metals in ground water of a portion of the front range mineral belt. OWRT Proj. A-023-Colo., Final Completion Rep. Environ. Resour. Center, Colorado State Univ., Fort Collins. 61 p.

Knollenberg, R. G. 1966a. Urea as an ice nucleant for supercooled clouds. J. Atmos. Sci. 23:197-201.

Knollenberg, R. G. 1966b. A note on a universal solid reagent seeding unit. J. Appl. Meteorol. 5:897-899.

Komabaysi, M., and Y. Ikebe. 1961. Organic ice nuclei: Ice forming properties of some aromatic compounds. J. Meteorol. Soc. Japan 39:82-94.

Kondratyev, T. S. 1971. Effect of silver water on some microorganisms. Farmatsevtichnii Zhurnal (Kiev) 26(1):62-65.

Kopp, J. F., and R. C. Kroner. 1970. Trace metals in waters of the United States. U.S. Dep. Interior Fed. Water Pollut. Control Admin., Cincinnati, Ohio. 218 p.

Kowalski, E., Z. Latallo, and S. Niewiarowski. 1956. Plasminogen and plasmin. II. Influence of metal ions on plasmin activity. Acta Biochem. Pol. 3:83-94.

Koyama, T., M. Hishida, and T. Tomino. 1970. Influence of sea salts on the soil metabolism. Soil Sci. Plant Nutr. 16:81-86.

Kratochvil, B., S. L. Boyer, and G. P. Hicks. 1967. Effect of metals on the activation and inhibition of isocitric dehydrogenase. Anal. Chem. 39:45-51.

Krauskopf, K. B. 1956. Factors controlling the concentrations of thirteen rare metals in sea-water. Geochim. Cosmochim. Acta 9:1-32b.

Kugelman, I. J., and K. K. Chin. 1971. Toxicity, synergism, and antagonism in anaerobic waste treatment processes. Advances in Chemistry Series 105, Am. Chem. Soc., Washington, D.C. p. 55-90.

Kugelman, I. J., and P. L. McCarty. 1965. Cation toxicity and stimulation in anaerobic waste treatment. I. Slug feed studies. J. Water Pollut. Control Fed. 37:97-115.

Kury, J. W., A. Z. Zielen, and W. M. Latimer. 1953. Heats of formation and entropies of HS^- and S^{2-}. J. Electrochem. Soc. 100:468-470.

Lande, E. 1977. Heavy metal pollution in Trondheimsfjorden, Norway, and the recorded effects on the fauna and flora. Environ. Pollut. 12:187-198.

Langer, G., C. Nagamoto, J. Rudgers, and J. Rosinski. 1974. Evaluation of the seeder for the NHRE suppression rocket. Proc. 4th Conf. Weather Modification. Am. Meteorol. Soc., Fort Lauderdale, Fla. p. 236-241.

Langer, G., J. Rosinski, and S. Bernsen. 1963. Organic crystals as ice nuclei. J. Atmos. Sci. 20:557-562.

Langer, G., J. Rosinski, and C. P. Edwards. 1967. A continuous ice nucleus counter and its application to tracking in the troposphere. J. Appl. Meteorol. 6:114-125.

Langmuir, I. 1948. The growth of particles in smokes and clouds and the production of snow from supercooled clouds. Proc. Am. Philos. Soc. 92:167-185.

La Riviere, J. W. M. 1977. Microbial ecology of liquid waste treatment. Adv. Microb. Ecol. 1:215-259.

Laskowski, N., T. Kost, D. Pommerenke, A. Schafer, and H. J. Tobschall. 1975. Heavy metal and organic carbon content of recent sediments near Mainz. Naturwissenschaften 62:136.

Latimer, W. M. 1952. The oxidation states of the elements and their potentials in aqueous solutions. 2nd ed. Prentice-Hall, Englewood Cliffs, N.J. 392 p.

Latimer, W. M., K. S. Pitzer, and W. V. Smith. 1938. The entropies of aqueous ions. J. Am. Chem. Soc. 60:1829-1831.

Lepp, N. W. 1975. The potential of tree-ring analysis for monitoring heavy metal pollution patterns. Environ. Pollut. 9:49-61.

Lewis, G. E., and M. Randall. 1961. Thermodynamics. 2nd ed. (Revised by K. S. Pitzer and L. Brewer.) 723 p.

Lietzke, M. H., and J. O. Hall. 1967. The distribution of positively charged cerium (III) and gadolinium complexes between a cation-exchanger and an aqueous phase. J. Inorg. Nucl. Chem. 29:2441-2452.

Lindow, S. E., D. C. Arny, W. R. Barchet, and C. D. Upper. 1975. The relationship between populations of bacteria active in ice nucleation and frost sensitivity in herbaceous plants. EOS Trans. Am. Geophys. Union 56:994.

Lindow, S. E., D. C. Arny, and C. D. Upper. 1976. *Erwinia herbicola* isolates active in ice nucleation inside frost damage to corn. Proc. Am. Phytopathol. Soc. 3:224.

Lindsay, W. L. 1978. Chemical equilibria in soils. John Wiley & Sons, New York (in press).

Lindsay, W. L., and M. Sadiq. 1978. Use of pe + pH as a redox parameter in soils. Soil Sci. Soc. Am. J., Vol. 42.

Linstedt, K. D., C. P. Houck, and J. T. O'Connor. 1971. Trace element removals in advanced wastewater treatment processes. J. Water Pollut. Control Fed. 43:1507-1513.

Luoma, S. N., and E. A. Jenne. 1975. The availability of sediment-bound cobalt, silver, and zinc to a deposit-feeding clam. Proc. 15th Life Sciences Symp. Biol. Implications of Metal in the Environment, Hanford, Wash. 33 p.

Luoma, S. N., and E. A. Jenne. 1976. Estimating bioavailability of sediment-bound trace metals with chemical extractants. 10th Annu. Conf. Trace Substances in Environ. Health, Columbia, Mo. 15 p.

McCabe, L. J. 1969. Trace metals content of drinking water from a large system. 157th Nat. Meeting Am. Chem. Soc., Minneapolis, Minn. (Abstr. WATR 013).

McCarty, P. L. 1971. Energetics and kinetics of anaerobic treatment. Advances in Chemistry Series 105, Am. Chem. Soc., Washington, D.C. p. 91-107.

McElroy, R. O., and D. W. Duncan. 1974. Copper extraction by rapid bacteriological process. U.S. Patent 3,856,913, Dec. 24, 1974. B. C. Res. Counc.

McIntosh, A. W. 1974. Notes on the use of copper sulfate in ponds. Bull. Environ. Contam. Toxicol. 12:425-432.

McIntosh, A. W. 1975. Fate of copper in ponds. Pestic. Monit. J. 8:225-231.

McIntyre, J. M., and E. S. Amis. 1968. Electromotive force of the silver-silver iodide, hydrogen iodide, hydrogen cell and the activity of hydrogen iodide in water-methanol media. J. Chem. Eng. Data 13:371-375.

McKee, J. E., and H. W. Wolf. 1963. Water quality criteria. 2nd ed. Calif. State Water Qual. Control Board Publ. 3A. 548 p.

Maki, L. R., E. Galyan, M. Chang-Chien, and D. R. Caldwell. 1974. Ice nucleation induced by *Pseudomonas syringae*. Appl. Microbiol. 28:456-459.

Maki, L. R., and D. M. Garvey. 1975. Bacterially induced ice nucleation. EOS Trans. Am. Geophys. Union 56:994.

Mangel, M. S. 1971. A treatment of complex ions in sea water. Am. Mar. Geol. 11:M24-M26.

Marsh, P. B., M. E. Simpson, and M. W. Trucksess. 1975. Effects of trace metals on the production of aflatoxins by *Aspergillus niger*. Appl. Microbiol. 30:52-57.

Martell, A. E., and M. Calvin. 1952. Chemistry of the metal chelate compounds. Prentice-Hall, Inc., New York. 613 p.

Martinek, K., K. H. Vill', Z. A. Strel'tsova, and I. V. Berezin. 1969. Structure of the α-chymotrypsin active site as revealed from a kinetic investigation of the inhibitory effect of silver ions. Mol. Biol. 3:554-565.

Martz, F. A., P. J. Van Soest, J. R. Vogt, and E. S. Hilderbrand. 1974. Use of elemental tracers and activation analysis in digestion, rate of ingesta flow, and food particle tracking studies in cattle. Eur. Assoc. for Anim. Prod. Publ. 14. p. 111-114.

Mason, B. 1966. Principles of geochemistry. 3rd ed. John Wiley & Sons, Inc., New York. 329 p.

Mason, B. J., and J. Hallett. 1956. Artificial ice forming nuclei. Nature (Lond.) 177:681.

Mason, B. J., and A. P. Van den Heuvel. 1959. Properties and behavior of some artificial ice nuclei. Proc. Phys. Soc. 74:744.

Matrone, G. 1969. Biochemistry and mechanisms of action of trace elements. Proc. 8th Int. Congr. Nutr. p. 171-175.

Maxfield, D., J. M. Rodriguez, M. Buettner, J. Davis, L. Forbes, R. Kovacs, W. Russel, L. Schultz, R. Smith, J. Stanton, and C. M. Wai. 1974a. Heavy metal content in sediments of southern part of Coeur d'Alene Lake. Environ. Pollut. 6:263-266.

Maxfield, D., J. M. Rodriguez, M. Buettner, J. Davis, L. Forbes, R. Kovacs, W. Russel, L. Schultz, R. Smith, J. Stanton, and C. M. Wai. 1974b. Heavy metal pollution in sediments of Coeur d'Alene River delta. Environ. Pollut. 7:1-6.

Merck Index. 1968. 8th ed. Paul G. Stecher (ed.) Merck and Co., Rahway, N.J. 1713 p.

Meskhi, A. B., and D. I. Dzhokhadze. 1973. Effects of simple phenols on activity of different systems. Sov. Plant Physiol. 20(6):1066-1069.

Miller, L. P., and S. E. A. McCallan. 1957. Toxic action of metal ions to fungus spores. Agric. Food Chem. 5:116-122.

Misaka, E., and A. L. Tappel. 1971. Inhibition studies of cathepsins A, B, C, and D from rat liver lysosomes. Comp. Biochem. Physiol. Ser. B 38:651-662.

Mitteldorf, A. J., and D. O. Landon. 1952. Spectrochemical determination of the mineral-element content of beef. Anal. Chem. 24:469-472.

Modak, S. M., and C. L. Fox. 1973. Binding of silver sulfadiazine to the cellular components of *Pseudomonas aeruginosa*. Biochem. Pharmacol. 22:2391-2404.

Molisch, H. 1918. Das Chlorophyllkorn als Reduktionsorgan. Sitzungsberichte. Akad. Wiss. Vienna 127:449-472.

Molise, E. M., and D. A. Klein. 1974. Effects of silver on radiorespiration of glucose by soil microorganisms. Annu. Meeting Am. Soc. Microbiol. (Abstr. E7).

Montie, T. C., and D. B. Montie. 1973. Selective detoxification of murine toxin from *Yersinia pestis*. Reaction of heavy metals with essential sulfhydryl and tryptophan residues. Biochemistry 12:4958-4965.

Mortimer, C. H. 1942. The exchange of dissolved substances between mud and water in lakes. J. Ecol. 30:147-201.

Mortvedt, J. J., and P. M. Giordano. 1975. Response of corn to zinc and chromium in municipal wastes applied to soil. J. Environ. Qual. 4:170-174.

Mossop, S. C. 1956. The nucleation of supercooled water by various chemicals. Proc. Phys. Soc., Ser. B 69:165.

Mossop, S. C., and C. Tuck-lee. 1968. The composition and size distribution of aerosols produced by burning solutions of AgI and NaI in acetone. J. Appl. Meteorol. 7:234-240.

Mount, D. I., and W. A. Brungs. 1967. A simplified dosing apparatus for fish toxicological studies. Water Res. 1:21-29.

Mulvey, G. J. 1977. Physical mechanisms of extra-area effects from weather modification. Dep. Atmos. Sci. Pap. No. 276, Colorado State Univ., Fort Collins. 138 p.

Mulvey, G. J., and L. O. Grant. 1976. A physical mechanism of extra-area effects from the Climax orographic cold cloud seeding experiment. Proc. 2nd WMO Sci. Conf. Weather Modification, Boulder, Colo. p. 473-479.

Mulvey, G. J., and J. Scheaffer. 1978. Evidence of long range transport of seeding material. J. Appl. Meteorol. (in preparation).

Murray, C. N., and L. Murray. 1973. Adsorption-desorption equilibria of some radionuclides in sediment-fresh-water and sediment-seawater systems, p. 105-124. *In* Proc. Radioactive Contaminants of the Marine Environment Symp., Int. Atomic Energy Agency, Seattle, Wash.

Mustafaev, F. M., and A. S. Abbasov. 1970. Thermodynamic properties of silver sulfide (Ag_2S). Ivd. Akad. Nauk. Azerb. SSR, Ser. Fiz-Tekh. Mat. Nauk (3):90-91.

Myrback, K. 1963. Studies on yeast invertase. XIV. The binding of silver ion by the enzyme. Arkiv Kemi 21:221-237.

Myrback, K., and E. Willstaedt. 1957. Inhibition of yeast invertase (saccharase) by metal ions. Arkiv Kemi 11:275-283.

Mytelka, A. I., J. S. Czachor, W. B. Guggino, and H. Golub. 1973. Heavy metals in waste water and treatment plant effluents. J. Water Pollut. Control Fed. 45(9):1859-1864.

Nageli, C. von. 1893. Uber die oligodynamischen Erscheinungen an lebenden Zellen. Neue Denkschriften der allg. Schweizer Gesellschaft fur die gesamte Naturwissenschaft Abt. 1.

Nakamura, S., and Y. Ogura. 1968. Mode of inhibition of glucose oxidase by metal ions. J. Biochem. (Tokyo) 64(4):439-447.

Nathanson, J. A., and F. E. Bloom. 1976. Heavy metals and adenosine cyclic 3',5'monophosphate metabolism: Possible relevance to heavy metal toxicity. Mol. Pharmacol. 12:390-398.

Natusch, D. F. S., J. R. Wallace, and C. A. Evans, Jr. 1974. Toxic trace elements: Preferential concentration in respirable particles. Science 183:202-204.

Naumov, G. B., B. N. Ryzhenko, and I. L. Khodakovsky. 1974. Handbook of thermodynamic data. MoscowAtomizdat. 1971. (Translated by G. J. Soleimani.) I. Barnes and V. Speltz (eds.) U.S. Geol. Surv. NTIS PB-226 722. 328 p.

Nehring, R. B. 1973. Heavy metal toxicity in two species of aquatic insects. M.S. thesis. Colorado State Univ., Fort Collins. 82 p.

Nielson, L. W., and L. M. Massey. 1940. Silver as a fungicide, p. 431-450. *In* L. Addicks (ed.) Silver in industry. Reinhold Publishing Co., New York.

Noddack, I., and W. Noddack. 1939. Die Haufigkeiten der Schwermetalle in Meerestieren. Arkiv Zool. 32A(1,4):1-35.

Nordstrom, D. K., and E. A. Jenne. 1977. Fluorite solubility equilibria in selected geothermal waters. Geochim. Cosmochim. Acta 41:175-188.

Novak, J. T., and M. S. Ramesh. 1975. Stimulation in anaerobic degradation. J. Water Res. 9:963-967.

Novello, F., and F. Stirpe. 1969. Effects of copper and other ions on the ribonucleic acid polymerase activity of isolated rat liver nuclei. Biochem. J. 111(1):115-119.

Obydennyi, P. T. 1976. Stimulating plant growth of trees and shrubs by treatment with a solution of a stimulant. Otkrytiya, Izobret., Prom. Obraztsy, Touarnye Znaki 53(26):6.

Olcott, C. T. 1950. Experimental argyrosis. V. Hypertrophy of the left ventricle of the heart in rats ingesting silver salts. Arch. Pathol. 49:138-149.

Oppenheimer, B. S., E. T. Oppenheimer, I. Danishefsky, and A. P. Stout. 1956. Carcinogenic effect of metals in rodents. Cancer Res. 16:439-441.

Paces, T. 1973. Steady-state kinetics and equilibrium between ground water and granitic rock. Geochim. Cosmochim. Acta 37:2641-2663.

Pankratz, L. B. 1970. Thermodynamic data for silver chloride and silver bromide. U.S. Bur. Mines Invest. Rep. 7430.

Parker, V. B. 1965. Thermal properties of aqueous univalent electrolytes. U.S. Nat. Bur. Standards Ref. Data 2:1-66.

Parungo, F. P., E. Ackerman, and R. F. Pueschel. 1974. AgI ice nuclei: Physical and chemical properties depending on their generating procedure. Proc. 4th Conf. Weather Modification. Am. Meteorol. Soc., Fort Lauderdale, Fla. p. 165-172.

Parungo, F. P., and J. P. Lodge. 1967. Amino acids as ice nucleators. J. Atmos. Sci. 22:274-277.

Parungo, F., and C. E. Robertson. 1969. Silver analysis of seeded snow by atomic absorption spectroscopy. J. Appl. Meteorol. 8:315-321.

Passerelli, R. E., Jr., H. Chessin, and B. Vonnegut. 1973. Ice nucleation by solid solutions of silver-copper iodide. Science 181:549-551.

Passerelli, R. E., Jr., H. Chessin, and B. Vonnegut. 1974. Ice nucleation in a supercooled cloud by CuI-3AgI and AgI aerosols. J. Appl. Meteorol. 13:946-948.

Patrick, R. 1973. Use of algae, especially diatoms, in the assessment of water quality, p. 76-95. *In* J. Cairns, Jr. and K. L. Dickson (eds.) Biological methods for the assessment of water quality. Am. Soc. Testing Materials Publ. 528.

Patrikeev, V. V., A. D. Malkina, and A. I. Kartsivadze. 1974. Crystallization of water in undercooled clouds and fogs. Fr. Demande 2:190-353.

Peacock, R. D. 1959. Potassium fluoroaurate (III). Chem. and Ind. 904.

Peterson, J. T., and C. E. Junge. 1971. Man's impact on the climate. MIT Press, Cambridge, Mass. 594 p.

Pettersson, O. 1976. Heavy-metal ion uptake by plants from nutrient solutions with metal ions, plant species, and growth period variations. Plant Soil 45:445-459.

Phalen, R. F., and P. E. Morrow. 1973. Experimental inhalation of metallic silver. Health Phys. 24:509.

Pine, M. J. 1971. The methane fermentations. Advances in Chemistry Series 105, Am. Chem. Soc., Washington, D.C. p. 1-10.

Pitcher, R. S., and D. G. McNamara. 1972. The toxicity of low concentrations of silver and cupric ions to the three species of plant-parasitic nematodes. Nematologica 18:385-390.

Pleskov, Y. V., and B. N. Kabanov. 1957. Composition of the complex silver ion in strongly alkaline solutions. (Cited by Sillen and Martell, 1964.) Shur. Neorg. Khim. 2:1807-1811.

Poon, C. P. C., and K. H. Bhayani. 1971. Metal toxicity to sewage organisms. J. Sanit. Eng. Div. Proc. Am. Soc. Civ. Eng. 97(SA2):161-169.

Popelka, M., N. Hejtmankova, I. Leifertova, J. Kurdnocova, and F. Santavy. 1973. Antifungal effects of some thymol derivatives and other substances. Acta Univ. Palacki. Olomuc., Fac. Med. 65:5-17.

Porter, J. R. 1946. Bacterial chemistry and physiology. John Wiley & Sons, Inc., New York. 1073 p.

Power, B. A., and R. F. Power. 1962. Some amino acids as ice nucleators. Nature (Lond.) 194:1170-1171.

Presant, E. W. 1963. A trace element study of some selected soil profiles from the Bathurst, N. B. area. M.S. thesis. Carleton Univ., Ottawa. (Cited in Boyle, 1968.)

Presant, E. W., and W. M. Tupper. 1965. Trace elements in some New Brunswick soils. Can. J. Soil Sci. 45:305-310.

Primack, M. P. 1975. Stimulation of mitochondrial protein synthesis by metal ions. Biochem. Biophys. Acta 407:439-441.

Pruppacher, H. R., and R. Sanger. 1955. Mechanismus der Vereinung unterkuhlter Wassertropfen durch disperse Keimsubstanzen. Z. angew. Math. Phys. 6:407.

Prusik, T., and N. E. Geacintov. 1976. Triplets of aromatic dyes bound to DNA as probes of the mobility of DNA-bound metal ions. FEBS (Fed. Eur. Biochem. Soc.) Lett. 71:236-240.

Ramette, R. W. 1972. Silver equilibrium via cell measurements. Quantitative laboratory experiments. J. Chem. Educ. 49:423-424.

Rattonetti, A. 1974. Determination of soluble cadmium, lead, silver, and indium in rain water and stream water with the use of flameless atomic absorption. Anal. Chem. 46:739-742.

Rayman, M. K., T. C. Y. Lo, and B. D. Sanwal. 1972. Transport of succinate in *Escherichia coli*. II. Characteristics of uptake and energy coupling with transport in membrane preparations. J. Biol. Chem. 247:6332-6339.

Reardon, E. J., and D. Langmuir. 1974. Thermodynamic properties of the ion pairs $MgCO_3$ and $CaCO_3$ from 10 to 15°C. Am. J. Sci. 274:599-612.

Reid, J. 1976. Dispersion in a mountain environment. Ph.D. dissertation. Dep. Atmos. Sci., Colorado State Univ., Fort Collins. 150 p.

Reinking, R. F., and L. O. Grant. 1968. The advection of artificial ice nuclei to mountain clouds from ground-based generators. Proc. 1st Conf. Weather Modification. Am. Meteorol. Soc., Boston, Mass. p. 433-445.

Richmond, R. C., and E. C. Powers. 1974. Modification of radiation sensitivity of bacterial spores by silver salts. Radiat. Res. 58:470-480.

Riley, J. P., and R. Chester. 1971. Introduction to marine chemistry. Academic Press, Inc., New York. 465 p.

Robie, R. A., and D. R. Waldbaum. 1968. Thermodynamic properties of minerals and related substances at 298.15°K (25°C) and one atmosphere (1.013 Bars) pressure and at higher temperatures. U.S. Geol. Surv. Bull. 1259. 256 p.

Rogers, K. S. 1969. Protein fluorescence and inactivation of glutamate dehydrogenases during treatment with silver. Enzymologica 36(3):153-160.

Romans, I. B. 1968. Oligodynamic metals, p. 372-400, *and* Silver compounds, p. 469-475. *In* C. A. Lawrence and S. S. Block (eds.) Disinfection, sterilization, and preservation. Lea and Febiger, Philadelphia, Penn.

Rosenquist, J. 1949. J. Metals 1:151. (Cited by Wagman and Kilday, 1973.)

Roth, W. A. 1932. The thermodynamic of the system sulfur dioxide-water. Z. Phys. Chem. A173:313-318.

Rottner, D., S. Brown, and O. Foehner. 1975. The effects of persistence of AgI on randomized weather modification experiments. J. Appl. Meteorol. 14:939-945.

Roy, D. R., and J. A. Bailey. 1974. Effect of silver from cloud seeding on rumen microbial function. Water, Air, Soil Pollut. 3:343-351.

Sabbioni, E., and F. Girardi. 1977. Metallobiochemistry of heavy metal pollution: Nuclear and radiochemical techniques for long-term, low level exposure (LLE) experiments. Sci. Total Environ. 7:145-179.

Sadiq, M., and W. L. Lindsay. 1978. Selection of standard free energies of formation for use in soil science. Colorado State Univ. Exp. Stn. Special Publ. (in press).

Sax, N. I. 1963. Silver-silver thioarsenite, p. 1174-1178. *In* Dangerous properties of industrial materials. 2nd ed. Reinhold Publishing Corp., New York.

Sax, N. I. 1975. Dangerous properties of industrial materials. 4th ed. Van Norstrand Reinhold Co., New York. 1258 p.

Sayre, J. D. 1952. Some hidden secrets of plant nutrition. Fertil. Rev. 28:8-9.

Schaefer, V. J. 1946. The production of ice crystals in a cloud of supercooled water droplets. Science 104:457-459.

Schaefer, V. J. 1976. After twenty-nine years—a proposal. J. Weather Modification 8:190-196.

Schaefer, V. J., and J. J. Fuquay. 1965. The detection of silver iodide in snow by neutron activation analysis. J. Rech. Atmos. 2:49-52.

Schmitt, J., H. Ritter, and K. Schmidt. 1975. Lactate dehydrogenase: Inhibition of submit A by the sulfhydryl reagent $AgNO_3$. Humangenetik 28:239-243.

Schnell, R. C. 1976. Bacteria acting as natural ice nucleants at temperatures approaching -1°C. Bull. Am. Meteorol. Soc. 57:1356-1357.

Schnell, R. C., J. F. Carney, and C. E. Carty. 1975. Ocean derived ice nuclei. Proc. Conf. Interactions of the Sea and the Atmosphere, Seattle, Wash.

Schnell, R. C., and G. Vali. 1972. Atmospheric ice nuclei from decomposing vegetation. Nature (Lond.) 236:163-165.

Schnell, R. C., and G. Vali. 1976. Biogenic ice nuclei. Part I. Terrestrial and marine sources. J. Atmos. Sci. 33:1554-1564.

Schroeder, H. A. 1960. Possible relationships between trace metals and chronic diseases, p. 59-67. *In* M. J. Seven and L. A. Johnson (eds.) Metal-binding in medicine. J. B. Lippincott Co., Philadelphia, Penn.

Scott, K., and J. G. Hamilton. 1950. Metabolism of silver in the rat with radiosilver as an indicator. Univ. Calif. Publ. Pharmacol. 2(19):241-262.

Scott, W. L., Jr. 1967. Silver uptake in brains of chronically gamma-irradiated rats: A study of neutron activation analysis. Radiat. Res. 31:522-528.

Serpolay, R. 1954. Action de certains liquides frigorigenes sur les meages surfondus. Bull. Obs Puy de Dome 4:99-102.

Serpolay, R. 1960. L'attenuation des broillards sur les aerodromes. Forces Aeriennes Francaise 16:35-66.

Serpolay, R. 1965. A ground based device for dispersal of supercooled fogs. Proc. Int. Conf. Cloud Physics. Am. Meteorol. Soc., Tokyo, Japan. p. 410-413.

Seuderling, Y. 1933. The continuous oligodynamic effect of elements on bacteria. Duodecim. Alb. (Chem. Abstr. 29:2992[9]).

Sheppard, J. C., and W. H. Funk. 1975. Trees as environmental sensors monitoring long-term heavy metal contamination of Spokane River, Idaho. Environ. Sci. Technol. 9:638-642.

Shimizu, F., Y. Shimizu, and K. Kumagai. 1976. Specific induction of herpes simplex virus by silver nitrate at low concentrations and biological activities of the inactivated virus. Antimicrob. Agents Chemother. 10:57-63.

Sillen, L. G., and A. E. Martell. 1964. Stability constants of metal-ion complexes. 2nd ed. Special Publ. No. 17, Chem. Soc., London.

Silvey, W. D., and G. A. Irwin. 1969. Relation of water quality to striped-bass mortalities in the Carquinez Strait of California. U.S. Geol. Surv. Open-File Rep. WRD. 12 p.

Simpson, H. J. 1976. Man-made radionuclides and sedimentation in the Hudson River estuary. Science 194:179-183.

Singleton, P., and R. White. 1975. Silver iodide: Its movement in an alpine ecosystem and its effect on two plant species, p. 329-351. *In* D. H. Knight (ed.) The Medicine Bow ecology project. Univ. Wyoming, Laramie.

Sinley, J. R., J. P. Goettl, Jr., and P. H. Davies. 1974. The effects of zinc on rainbow trout (*Salmo gairdneri*) in hard and soft water. Bull. Environ. Contam. Toxicol. 12(2):193-201.

Skei, J. M., N. B. Price, S. E. Calvert, and H. Holtedahl. 1972. The distribution of heavy metals in sediments of Sorfjord, West Norway. Water, Air, Soil Pollut. 1:452-461.

Smith, I. C., and B. L. Carson. 1977. Trace metals in the environment. Vol. II. Silver. Ann Arbor Science Publ., Inc. 490 p.

Smith, R. G., Jr., and H. L. Windom. 1972. Analytical handbook for the determination of As, Cd, Co, Cu, Fe, Pb, Mn, Hg, Ni, Ag and Zn in the marine and estuarine environments. Mar. Resour. Center, Savannah, Ga. 62 p.

Snodgrass, P. J., B. L. Vallee, and F. L. Hoch. 1960. Effects of silver and mercurials on yeast alcohol dehydrogenase. J. Biol. Chem. 235:504-508.

Sokol, R. A., and D. A. Klein. 1975. The response of soils and soil microorganisms to silver iodide weather modification agents. J. Environ. Qual. 4:211-214.

Somers, E. 1959. Plant pathology: Fungitoxicity of metal ions. Nature (Lond.) 184:475-476.

Spadaro, J. A., and R. O. Becker. 1976. Some specific effects of electrically injected silver and gold ions. Bioelectrochem. Bioenerg. 3:49-57.

Speece, R. E., and P. L. McCarty. 1964. Nutrient requirements and biological solids accumulation in anaerobic digestion, p. 305-333. *In* W. W. Eckenfelder (ed.) Advances in Water Pollution Research, Vol. 2. The Macmillan Co., New York.

Sprague, J. B. 1969. Measurement of pollutant toxicity to fish. I. Bioassay methods for acute toxicity. Water Res. 3:793-821.

Standler, R. B., and B. Vonnegut. 1972. Estimated possible effects of AgI cloud seeding on human health. J. Appl. Meteorol. 11(8):1388-1391.

Steffe, D. 1970. Concentration variation of some trace materials in precipitation from Great Plains thunderstorms. Precipitation Scavenging, AEC Symp. Ser. 22. p. 109-120.

Stehlik, B. 1963. Oxidation-reduction potential of Ag_2O_3 in acidic silver salts. Chem. Szesti. 17:6-13.

Stern, E. 1974. Portable water purifier (Dynek Corp.) German offen 2,357,585. 6 June 1974 (Chem. Abstr. 81:P96312z).

Sternlieb, I., and S. Goldfischer. 1968. Heavy metals and lysosomes. Front. Biol. 45:185-200.

Steven, S. E. 1977. Aeromonas: Physiologic changes during starvation and secondary stress. M.S. thesis. Colorado State Univ., Fort Collins. 85 p.

Stotzky, G. 1965. Microbial respiration, p. 1556-1572. *In* C. A. Black (ed.) Methods of soil analysis. Part 2. Chemical and microbiological properties. Am. Soc. Agron., Madison, Wisc.

Struempler, A. 1976. Trace metals in rain and snow during 1973 at Chadron, Nebraska. Atmos. Environ. 10:33-37.

Stumm, W., and H. Bilinski. 1973. Trace metals in natural waters; difficulties of interpretation arising from our ignorance on their speciation, p. 39-52. *In* S. H. Jenkins (ed.) Advances in water pollution research. Pergamon Press, New York.

Summers, A. O., and L. I. Sugarman. 1974. Cell-free mercury(II)-reducing activity in a plasmid-bearing strain of *E. coli*. J. Bacteriol. 119(1):242-249.

Summers, P. 1972. The silver fallout patterns in precipitation from seeded convective storms. Proc. 3rd Conf. Weather Modification. Am. Meteorol. Soc., Rapid City, S.D. p. 279-286.

Super, A. B., E. W. Holroyd, III, and B. A. Silverman. 1977. The practicability of dry ice for on-top seeding of convective clouds. Proc. 6th Conf. on Planned and Inadvertant Weather Modification. Am. Meteorol. Soc., Urbana, Ill. p. 358-361.

Suzuki, N. 1958. Conditions for sampling for quantitative determination of traces of cobalt, zinc and silver in plants utilizing radioactive isotopes. Sci. Rep. Res. Inst. Tohoku Univ. 42:149-157.

Swanson, V. E., I. C. Frost, L. F. Rader, Jr., and C. Huffman, Jr. 1966. Metal sorption by northwest Florida humate. U.S. Geol. Surv. Prof. Pap. 550:C174-C177.

Tabatabai, M. A. 1977. Effects of trace elements on urease activity in soils. Soil Biol. Biochem. 9:9-13.

Takimoto, A., and O. Tanaka. 1973. The effects of some SH-inhibitors and EDTA on flowering in *Lemna perpusilla*. Plant Cell Physiol. 14:1133-1141.

Taylor, F. B. 1971. Trace elements and compounds in waters. J. Am. Water Works Assoc. 63(11):728-733.

Teller, H. L., D. R. Cameron, and D. A. Klein. 1976. Deposition of silver iodide used as a seeding agent in ecological impacts of snowpack augmentation in the San Juan Mountains of Colorado, p. 105-139. *In* Final report of San Juan Ecology project. U.S. Dep. Commerce, NTIS PB-255-012.

Teller, H. L., and D. A. Klein. 1973. Disposition and environmental impact of silver iodide in the National Hail Research Experiment. Operations Rep. No. 2. Colorado State Univ., Fort Collins. 88 p. + App.

Teller, H. L., and D. A. Klein. 1974. Disposition and environmental impact of silver iodide in the National Hail Research Experiment. Operations Rep. No. 3. Colorado State Univ., Fort Collins. 105 p. + App.

Terhaar, C. J., W. S. Ewell, S. P. Dziuba, and D. W. Fassett. 1972. Toxicity of photographic processing chemicals to fish. Photogr. Sci. Eng. 16:370-377.

Thurberg, F. P., A. Calabrese, and M. A. Dawson. 1974. Effects of silver on oxygen consumption of bivalves at various salinities, p. 67-78. *In* F. J. Vernberg and W. B. Vernberg (eds.) Pollution and physiology of marine organisms. Academic Press, Inc., New York.

Thurberg, F. P., and R. S. Collier. 1977. Respiratory response of cunners to silver. Mar. Pollut. Bull. 8:40-41.

Tiffin, L. O. 1975. The form and distribution of metals in plants: An overview. 15th Annu. Hanford Life Sci. Symp., Biol. Implications of Metals in the Environ., Richland, Wash. 35 p.

Tipton, I. H., and M. J. Cook. 1963. Trace elements in human tissue. Part II. Adult subjects from the United States. Health Phys. 9:103-145.

Tipton, I. H., P. L. Stewart, and P. G. Martin. 1966. Trace elements in diets and excreta. Health Phys. 12:1683-1689.

Truesdell, A. H., and B. F. Jones. 1973. WATEQ, a computer program for calculating chemical equilibria of natural waters: PB2-20464. 77 p.

Truesdell, A. H., and B. F. Jones. 1974. WATEQ, a computer program for calculating chemical equilibria of natural waters. J. Res. U.S. Geol. Surv. 2(2):233-274.

Tsai, C. G., and G. A. Jones. 1975. Isolation and identification of rumen bacteria capable of anaerobic phloroglucinol degradation. Can. J. Microbiol. 21(6):794-801.

Turekian, K. K., and M. R. Scott. 1967. Concentrations of Cr, Ag, Mo, Ni, Co, and Mn in suspended material in streams. Environ. Sci. Technol. 1:940-942.

Udall, N. D. 1973. Toxicity of the molluscides metaldehyde and methiocarb to dogs. Vet. Rec. 93(15):420-422.

U.S. Geological Survey. 1968. Studies related to wilderness-primitive areas: Uncompahgre, Colorado. U.S. Geol. Surv. Bull. 1261-C.

Vali, G., M. Christensen, R. W. Fresh, E. L. Galyan, L. R. Maki, and R. C. Schnell. 1976. Biogenic ice nuclei. Part II. Bacterial sources. J. Atmos. Sci. 33:1565-1570.

Vallee, B. L., and D. D. Ulmer. 1972. Biochemical effects of mercury, cadmium and lead. Annu. Rev. Biochem. 41:91-128.

Van Campen, D. R. 1966. Effects of zinc, cadmium, silver and mercury on the absorption and distribution of copper-64 in rats. J. Nutr. 88(1):125-130.

Vanderzee, C. E., and A. S. Quist. 1961. The third dissociation constant of orthophosphoric acid. J. Phys. Chem. 65:118-123.

Van Loon, J. C., J. Lichwa, D. Ruttan, and J. Kinrade. 1973. The determination of heavy metals in domestic sewage treatment plant wastes. Water, Air, Soil Pollut. 2:473-482.

Veeh, H. H., and K. K. Turekian. 1968. Cobalt, silver, and uranium concentration of reef-building corals in the Pacific Ocean. Limnol. Oceanogr. 13:304-308.

Verschuuren, H. G., R. Krodes, E. M. Dentonkelaar, J. M. Berkvens, P. W. Helleman, and G. J. Van Esch. 1975. Long term toxicity and reproduction studies with metaldehyde in rats. Toxicology 4(1):97-115.

Vinogradov, A. P. 1953. The elementary chemical composition of marine organisms. Memoir No. II. Sears Found. Mar. Res., Yale Univ., New Haven, Conn. 587 p.

Vonnegut, B. 1947. The nucleation of ice formation by silver iodide. J. Appl. Phys. 18:593-595.

Vonnegut, B. 1948. Production of ice crystals by adiabatic expansion of gas. J. Appl. Phys. 19:959.

Vonnegut, B., and H. Chessin. 1971. Ice nucleation by coprecipitated silver iodide and silver bromide. Science 174:945-946.

Vosburgh, W. C., and R. S. McClure. 1943. Complex ions. IV. Monoamine silver ion. J. Am. Chem. Soc. 65:1060-1063.

Waggoner, P. E. 1966. Weather modification and the living environment, p. 87-98. *In* F. F. Darling and J. P. Milton (eds.) Future environments of North America. The Natural History Press, Garden City, N.Y.

Wagman, D. D., W. H. Evans, V. B. Parker, I. Halow, S. M. Bailey, and R. H. Schumm. 1969. Selected values of chemical thermodynamic properties. U.S. Nat. Bur. Std. Tech. Note 270-274.

Wagman, D. D., and M. V. Kilday. 1973. Enthalpies of precipitation of silver halides; entropy of silver ion. J. Res. Nat. Bur. Std. 77:569-579.

Wahlberg, J. E. 1965. Percutaneous toxicity of metal compounds. Arch. Environ. Health 11:201-204.

Wallace, A. 1971. Uptake and distribution of ^{185}W and ^{110}Ag in plants, p. 36-39. *In* A. Wallace (ed.) Regulation of the micronutrient status of plants by chelating agents and other factors. Edward Brothers, Inc., Ann Arbor, Mich.

Warburton, J. A. 1971. Physical evidence of transport of cloud-seeding materials into areas outside primary targets. Proc. Int. Conf. Weather Modification. Am. Meteorol. Soc., Canberra, Australia. p. 185-190.

Warburton, J. A. 1973. The distribution of silver in precipitation from two seeded Alberta hail storms. J. Appl. Meteorol. 12:677-682.

Warburton, J. A., G. Linkletter, M. Owens, L. Young, and A. Anderson. 1975. Measurement of silver in snow and firn from Siple, Byrd, and South Pole stations, Antarctica. J. Rech. Atmos. 9:119-126.

Warburton, J. A., and L. G. Young. 1968a. Neutron activation measurements of silver in precipitation from locations in western North America. J. Appl. Meteorol. 7:444-448.

Warburton, J. A., and L. Young. 1968b. Neutron activation procedures for silver analysis in precipitation. J. Appl. Meteorol. 7:433-443.

Warburton, J. A., and L. G. Young. 1972. Determination of silver in precipitations down to 10^{-11} M concentrations by ion exchange and neutron activation analysis. Anal. Chem. 44:2043-2045.

Warnick, S. L., and H. L. Bell. 1969. The acute toxicity of some heavy metals to different species of aquatic insects. J. Water Pollut. Control Fed. 41:280-284.

Weaver, J. N., A. Hanson, J. McGaughy, and F. K. Steinkruger. 1974. Determination of heavy metals in municipal sewage sludge by NAA. Water, Air, Soil Pollut. 3:327-335.

Weaver, T. W., III. 1971. Ecological effects of weather modification: Silver holding capacity of Montana soils. Proc. Mont. Acad. Sci. 31:40-46.

Weaver, T. W., III. 1972. Effects of seeding agents on microorganisms and plants, p. 34-40. *In* Impacts of induced rainfall on the Great Plains of Montana. Interim Res. Rep. No. 26, Mont. Agric. Exp. Stn., Montana State Univ., Bozeman.

Weaver, T. W., III, and D. Klarich. 1973. Ecological effects of silver iodide in terrestrial ecosystems: A preliminary study, Section 3. *In* Impacts of induced rainfall on the Great Plains of Montana. Final Res. Rep. No. 42, Mont. Agric. Exp. Stn., Montana State Univ., Bozeman. 18 p.

Weier, E. 1938. Factors affecting the reduction of silver nitrate by chloroplasts. Am. Bot. 25:501-507.

Weinstein, A. I., and J. R. Hicks. 1976. Use of compressed air for supercooled fog dispersal. J. Appl. Meteorol. 15:1226-1231.

Weinstein, A. I., and B. A. Silverman. 1972. Warm fog dispersal by airborne wide area hygroscopic particle seeding. Proc. 3rd Conf. Weather Modification. Am. Meteorol. Soc., Rapid City, S.D. p. 61-65.

Wester, P. O. 1972. Trace elements in RNA from beef heart tissue. Sci. Total Environ. 1:97-103.

Whanger, P. D., and P. H. Weswig. 1970. Effect of some copper antagonists on induction of ceruloplasmin in the rat. J. Nutr. 100:341-348.

Wheatland, A. B. 1954. Factors affecting the formation and oxidation of sulfides in a polluted estuary. J. Hyg. 52:194-210.

White, J., and D. J. Munns. 1951. Inhibitory effect of common elements towards yeast growth. J. Inst. Brewing 57(48 N.S.):175-179.

Whittaker, R. H. 1967. Ecological implications of weather modification, p. 367-384. *In* R. H. Shaw (ed.) Ground level climatology. Am. Assoc. Adv. Sci. Publ. 86, Washington, D.C.

Wildung, R. E., H. Drucker, T. R. Garland, and R. A. Pelroy. 1976. Transformation of trace metals by microorganisms. Abstr. Annu. Meeting Am. Soc. Agron. p. 141.

Wisniewski, J. 1976. Variability of rainwater silver concentrations in south Florida. 2nd WMO Sci. Conf. Weather Modification. Am. Meteorol. Soc., Boulder, Colo. p. 563-568.

Wolfe, R. S. 1971. Microbial formation of methane. Adv. Microb. Physiol. 6:107-146.

Woodriff, R., B. Culver, D. Shrader, and A. Super. 1973. Determination of subnanogram quantities of silver in snow by furnace atomic absorption spectrometry. Anal. Chem. 45:230-234.

Woodward, R. L. 1963. Review of the bactericidal effectiveness of silver. J. Am. Water Works Assoc. 55:881-888.

Wuhrmann, K., and F. Zobrist. 1958. Untersuchungen uber die Bakterizide Wirkung von Silber in Wasser. Schweiz. Z. Hydrol. 20:218-254.

Wysor, M. S. 1975a. Orally administered silver sulfadiazine: Chemotherapy and toxicology in CF-1 mice, *Plasmodium berghei* (malaria) and *Pseudomonas aeruginosa*. Chemotherapy 21(5):302-310.

Wysor, M. S. 1975b. Antimalarials: Screening of (orally administered) silver sulfonamides against *Plasmodium berghei*. Chemotherapy 21(5):280-284.

Yoshimura, S., and T. Wada. 1938. Hydrogen sulphide and ferrous sulphide in the bottom deposits of the brackish water lake, Hamanako, Sizuoka Prefecture. Proc. Imperial Acad. Ueno Park, Tokyo 14(3):115-120.

Young, R. G., and D. J. Lisk. 1972. The effect of copper and silver ions on algae. J. Water Pollut. Control Fed. 44(8):1643-1647.

Zemansky, G. M. 1974. Removal of trace metals during conventional water treatment. J. Am. Water Works Assoc. 66:606-609.

Zettlemoyer, A. C., N. Tcheurekdjian, and C. L. Hosler. 1963. Ice nucleation by hydrophobic substrates. Z. angew. Math. Phys. 14:496-502.

Zimmermann, W. 1952. Oligodynamische Silberwirkung. I. Uber den Wirkungsmechanismus. Z. f. Hygiene 135:403-420.

Index

Numbers in **bold face** type refer to tables or graphs.
Numbers in *italic* type refer to chapter summaries.